CULTURAL PERSPECTIVES ON THE MATHEMATICS CLASSROOM

Mathematics Education Library

VOLUME 14

The titles published in this series are listed at the end of this volume.

CULTURAL PERSPECTIVES ON THE MATHEMATICS CLASSROOM

Edited by

STEPHEN LERMAN
South Bank University, London

SPRINGER-SCIENCE+BUSINESS MEDIA, B.V.

A C.I.P. Catalogue record for this book is available from the Library of Congress.

ISBN 978-90-481-4424-2 ISBN 978-94-017-1199-9 (eBook)
DOI 10.1007/978-94-017-1199-9

Printed on acid-free paper

TABLE OF CONTENTS

INTRODUCTION

We live in culturally conscious times. It confronts us in diverse forms: debates on late-night British television on 'high' culture versus 'low' culture, T.S. Eliot versus Bob Dylan; the same-tasting food that can be eaten in the same packaging in the same surroundings from New York to Beijing to Harare; architecture that draws in features which step across time and distance and that reflects the alienation, not in its modernist form of the self from society, but of the self from the self; the fin de siècle search for identity through aggressive assertions of ever more fragmented nationalities; and the increasing market-orientation of schools and universities in Britain and other countries, forced to compete with each other for 'customers' who were once called students. Amongst the young, culture manifests itself in the elaborate semiology whereby, for instance, the way that one's 'trainers' are tied, if they are the right ones in the first place, can identify or exclude from the desired group.

Culture is invasive, both outwards around the globe and inwards into the construction of individual subjectivities. The apparently irresistible appetite of late twentieth century capitalism creates demands for its products, rather than merely seeking markets, and as a consequence we become the consumers that multi-national companies need to feed upon in order to continue to grow.

This preamble cannot circumvent the inevitable demand for a definition of what is meant by culture. I do not believe that there is one which is entirely satisfactory. Raymond Williams, who spent much of his life addressing precisely this subject found it "one of the most complex words in the English language". It is probably too easy in such circumstances to fall back on a loose interpretation of Wittgenstein's aphorism and say that "meaning is use", although the reader may well have concluded that in the early nineties the most convincing indication of the meaning of the word culture has become its over-use!

However, I will echo Raymond Williams and say that culture is ordinary. It is something we all possess and which in turn possesses us. It is not an exotic thing possessed by others who are different from 'us'. Many aspects of our culture are transparent to our gaze until we are confronted with cultural differences in others and so the potential for conflict in varying degrees arises. As someone said, the last thing to discover water is the fish!

Until recently, the mathematics classroom was seen as an homogeneous cultural entity, a place for learning an academic discipline, peopled by students who learn and do things properly, plus the deviants (the majority?) who don't. Where this view has come under scrutiny and critique, it can perhaps best be seen as a disillusionment with 'nature', that is with attempts to describe the nature of the individual and her/his intellectual development and subsequently prescribe, whether through platonism, behaviourism, or Piagetian/Bloomian structuralism. In place of the search for nature in the 'psychology' of the individual, social accounts of mathematics learning have proliferated; Bishop, Bauersfeld and Brown are some of those authors. At the same time one sees a parallel manifestation in mathematics itself, in Wittgenstein, Davis, Hersh, Bloor and Restivo.

Culture has become identified as the key to understanding the ground on which the learner appropriates meaning. But we would be deluding ourselves if we thought that it is possible or desirable to attempt to specify, identify, or exhaustively characterise that culture or those cultures and so predict those meanings. Such determinism is a search in the other direction. Instead one can view attempts at shifting the locus of mathematics learning to the social as an 'opening up' of the activities of the mathematics classroom. For instance, developments in multicultural and anti-racist mathematics, far from giving answers, enable questions to be asked: why are there imbalances in average earnings and infant mortality across the world as well as within cities; how is unemployment calculated; who gains from price increases; what are the major factors in the ethnic make-up of inner cities; which power structures are maintained by particular stereotypes?

Second, recent histories of mathematics, such as those by Bernal and Joseph, offer alternative stories of the development of mathematical ideas at the same time as deconstructing notions of 'history' as singular and true. It also opens up the possibility of other stories of equal value (Bloor).

Third, the appearance of mathematics viewed as a sociological phenomenon engenders evaluations of the function it serves in different societies and within societies for different groups, the status it endows, and its relationship to other kinds of knowledge (Douglas, Restivo). Within education it raises questions of whose mathematics forms the school curriculum, why failure is 'required', and why mathematics is held in such low regard by so many, even feared and hated.

How and where has the locus of learning shifted? In attempting to sketch out an answer to my own question, I find myself slipping from the word 'culture' to 'social'. These terms are not interchangeable but nor are they simply separated. One would perhaps think of gender stereotypes as cultural, but of 'gender' as socially constructed. One would talk of the culture of the community of mathematicians, treating it as monolithic for a moment, but one would also talk, for example, of the social outcomes of being a member of that cultural group. In terms of perspectives on learning, one could argue that a social perspective, at least in addition and perhaps in contrast to, a cognitive perspective opens the possibility for cultural analyses of the mathematics classroom. Such analyses begin with social groups, ideologies, status, identities, language and so on. In giving a brief reconstruction of aspects of the history of mathematics education in common to some countries, I will describe a developing focus on the social. Anthropology and other cultural studies are both contributors to that shift of focus and are stimulated and developed by it.

Events in academic discourses on the boundaries of mathematics education conspired, in the 1950's, to bring about a shift from a straightforward transmission view of teaching and learning. Linguistic analysis led to the characterisation of 'teaching' as an activity that could be said to be taking place if, and only if, learning was taking place. At the same time, Piaget, via Montessori and other influences, proposed that children actively construct knowledge and that learning is dominated by development. So the classroom became 'child-centred'; a regime of truth (Foucault, Walkerdine) where right behaviour which demonstrated the appropriate development was the goal and other behaviours were interpreted as Other, as deficiencies. The New Maths changed little, except to make the structure of the mathematics curriculum parallel the logical structure of mathematics as constructed by Bourbaki. Investigative learning in mathematics began to move the control for learning away from the teacher, and a new metaphor was required. Constructivism, the neo-Piagetian position (von Glasersfeld, Steffe, Cobb, Confrey) which focused even more strongly on the cognizing individual, offered this metaphor. Now, it could be said that we are all constructivists. In one version or another, the story is that of each child actively constructing her or his own knowledge. There are constructivist teachers, constructivist classrooms, and constructivist textbooks, whatever these might be. But the locus of learning and knowledge-construction has always been with the child, and knowledge itself is situated in the social; indeed how could it be otherwise? Studies of mathematical activities and concepts in other cultures, both in a cross-cultural sense (Bishop, Saxe, Pinxten) and in an intra-cultural sense (Scribner, Lave, Carraher) challenge the centrality of the autonomous individual. Thus new questions are generated; how does intersubjective, social knowledge become the individual's; how does cultural knowledge become the individual's? The focus, if not the locus, can be said to be shifting again, and is now on the social settings in which knowledge is created, and the meanings carried by those practices for the individual; the intersubjective to the intrasubjective (Vygotsky).

The shift to a social locus of mathematics learning has complicated notions of teaching and learning mathematics, but this merely reflects the complexity of the interrelated components: children; a teacher; knowledge; the school, and society. Simple answers exclude; defining who succeeds, and why, is to pathologise the Other. Where we would have wondered about improving the styles of textbooks or whether there should be textbooks at all (e.g. Nuffield mathematics), now our notion of 'text' has shifted and every mediation is text. Where we would have aimed for child-centred teaching styles we now examine learning styles and wonder how to acknowledge and celebrate their multiplicity. Where we would have wondered how to improve the performance of underachieving groups we now examine how those group identities are constructed in society, the home and the classroom - but still wonder what to do. Where we would research how to convey mathematical meaning we now examine how meanings shift between and within practices, including that of academic mathematics. Where generalisability in research meant aiming at describing the majority, we now deconstruct the notion of 'generalisable'; is it itself generalisable?

The chapters in this book attempt to engage with this complexity and offer original contributions to the ways in which one can interpret what happens in the mathematics classroom and ultimately facilitate children's learning of mathematics. They adopt theoretical perspectives whilst drawing on the classroom as both the source of investigation and the site of potential change and development. No collection can possibly be exhaustive of the theme; this one reflects the choices of the editor and the ideas of the individual writers. They have not been asked to support any particular argument. They have been invited, as workers with words as well as with people, to offer their thoughts on "Cultural Perspectives on the Mathematics Classroom".

Any attempt to classify the chapters into sections would be too narrow as it would fail to do justice to the many dimensions upon which they all draw. All that has been done is to group together those whose gaze begins from outside the classroom, focusing on a political (Richard Noss, John Volmink), an anthropological (Rik Pinxten) or a sociological (Leone Burton) critique, and those whose gaze begins from within the classroom, on the teacher (Linda Gattuso, Candia Morgan), the classroom language (David Pimm, Jeff Evans), or the social interactions (Stephen Lerman, Terry Wood). One chapter stands alone, in its attempt to provide an overview of the whole domain, identify research trends, and look for future directions (Marilyn Nickson).

In her chapter, the first in the book, Marilyn Nickson calls for new research methods to "study the reality of the class" and warns of the dangers of neglecting pupils' perspectives in a focus on changing teachers' expectations. Richard Noss (Chapter 2) looks at the social functioning of schooling in Western society and examines how political ideologies influence the mathematics curriculum. John Volmink (Chapter 3) looks critically at the rhetoric of 'mathematics for all' from the perspective of South Africa. He suggests that only schools can bring about a breakdown of the elitism currently inherent in mathematical practice. Leone Burton (Chapter 4) raises the issue of what might be meant by the culture of mathematics and questions whose culture it is. In particular she suggests that school mathematics is a reflection of the values of white western middle-class males and that this needs to be subjected to examination and critique. Rik Pinxten (Chapter 5) discusses what light anthropology can shed on the mathematics classroom. He argues that one can see relevance in 4 areas in particular, "types of learning, cognitive content, language structure and institutional aspects of teaching".

Chapter 6 begins the gaze from within the classroom. Linda Gattuso examines the relationship between the teacher's beliefs and actions and the learning of the students. She takes the novel approach of researching her own practice for the ways in which what she wishes to do as a teacher is affected by constraints such as class control, her 'mood' and so on. Candia Morgan (Chapter 7) looks at how the introduction of computers into the mathematics classroom can lead to the raising of issues which challenge teachers' beliefs about the teaching and learning of mathematics. David Pimm (Chapter 8) raises the question of how different kinds of teacher talk control and frame the classroom discourse. Terry Wood (Chapter 9) looks at the culture of an inquiry mathematics classroom, drawing on her work and that of her colleagues at Purdue University. She ends by discussing how some aspects of their work sheds light on debates between Piagetian or Vygotskian views of collaborative learning. Jeff Evans and Anna Tsatsaroni (Chapter 10) examine the nature of discursive practices as they position individuals in the mathematics classroom and offer some thoughts on the implications of their work for the teacher. Finally Stephen Lerman (Chapter 11) offers an analysis of the potential of a shift in focus from an individualistic psychological description of the learner to a social interaction setting for classroom mathematics learning.

One chapter has appeared elsewhere, and I acknowledge the publishers NCTM and the editor Doug Grouws for permission to print Marilyn Nickson's paper from the book "Handbook of Research on Mathematics Teaching and Learning", published by Macmillan, 1992, although the version published here has some modifications. Acknowledgements also to Cambridge University Press for permission to reprint pages from their school textbooks in Chapter 11.

I wish to acknowledge the help and support of my colleagues Rosalinde Scott-Hodgetts and David Blundell, whose early work helped this collection get off the ground, but who were unable to continue their involvement. This introduction draws on some of David's thoughts and ideas, but the remaining editor is alone responsible for the final outcome.

M. NICKSON

THE CULTURE OF THE MATHEMATICS CLASSROOM: AN UNKNOWN QUANTITY?[1]

The whole notion of a `culture of the mathematics classroom' implies an acceptance of the idea that mathematics exerts a unique influence on the context of classrooms in which the subject is being taught and learned. This is a relatively recent departure from traditional concerns in mathematics education and is one of the outcomes of shifts in perspectives within the field that have, in turn, given rise to some changes of emphasis in related research. A book such as this is in itself indicative of the importance given to new considerations of this kind. The aim of this chapter is to provide a theoretical setting for the chapters that follow which are concerned, in one way or another, with issues related to this change. In attempting to provide this setting, changes will be viewed from three broad perspectives: (a) those related to the nature of mathematics as a discipline; (b) those concerned with research about teachers and the teaching of mathematics, and (c) those concerning pupil perspectives.

This chapter will refer only to aspects of culture within the microcosm of the mathematics classroom in order to build a picture of what contributes to the establishment of a culture at this level.. The much wider issues that relate to mathematics education at the macro level will not be discussed. The choice to restrict this examination to the classroom alone is a deliberate one. There is a tendency in educational research to adopt global perspectives and popular catchwords without taking the time to clarify what we mean by them and to consider their relevance more thoroughly. Culture is just such a phenomenon. It appears increasingly in considerations at broader levels as, for example, with respect to ethnicity (D'Ambrosio, 1985), politics (Fasheh, 1982; Joseph, 1990; Mellin-Olsen, 1987; Pimm, 1990), technology (Noss, 1988), cross-cultural issues (Zaslavsky, 1989), and multicultural mathematics (Nickson, 1988b). If, however, we are concerned with the classroom as our unit of study, we are forced to come to some understanding and agreement about the meaning of culture at that level rather than taking for granted that the meaning is already accepted and shared. Once this is achieved, it becomes more meaningful to go beyond the classroom to consider wider aspects of culture and to explore, in turn, how they impinge on that of the mathematics classroom. This chapter is concerned with taking

---

[1]The author gratefully acknowledges the helpful comments provided by Peter Hall, University of Missouri, and Robert Underhill, Virginia Tech University.

S. Lerman (ed.), Cultural Perspectives on the Mathematics Classroom, 7–35.

the first steps towards understanding what makes up the culture of the mathematics classroom as a basis from which to view its place as one culture within a wider culture and so, hopefully lead to a better understanding of both.

WHAT IS THE CULTURE OF THE MATHEMATICS CLASSROOM?

A reasonably accessible view of culture from the wealth of interpretations available is offered by Levitas (1974) when he states:

> Every child in every society has to learn from adults the meanings given to life by his society; but every society possesses with a greater or lesser degree of difference, meanings to be learned. In short, every society has a culture to be learned though cultures are different. (p. 3)

Socialisation is seen as a universal process, and culture is seen as the content of the socialisation process that differs from one society to another. The shared meanings that come to be accepted by a society form its content. In discussing culture in the context of the school curriculum, Smith, Stanley, and Shores (1971) note that, in curricula as a whole, emphasis tends to be upon the "more fundamental universals, or cultural core, such as the values, sentiments, knowledges, and skills that provide society with stability and vitality and individuals with the motivations and deeplying controls of conduct" (p. 17). This is echoed by Feiman-Nemser and Floden (1986) when they state that a "focus on *culture* implies inferences about knowledge, values, and norms for action, none of which can be directly observed" (p. 506). In reviewing research related to cultures of teaching, they note that the idea of culture in this context offers a new way of examining educational practice suggestive of an anthropological perspective, although, as they point out, neither their approach nor the research they report is necessarily anthropological in nature.

For our purposes, what we shall be considering here are the invisible and apparently shared meanings that teachers and pupils bring to the mathematics classroom and that govern their interaction in it. While a single chapter cannot deal with all aspects of the knowledge, beliefs, and values that are held by the actors in this setting, by exploring the kinds of questions addressed by research in mathematics education we can identify where and how some of the matters of a cultural nature affect what goes on in the classroom.

A Variety of Classroom Cultures

There is a danger in associating culture with the mathematics classroom and that is to assume there is only one such culture. However, since key aspects of culture are concerned with unseen beliefs and values, the culture of a mathematics classroom will depend to a very large extent on these hidden perspectives of teachers and pupils in relation to the subject. As a result, just as there is a multiplicity of teacher cultures (Feiman-Nemser & Flodin 1986), there are many variations in the culture of mathematics classrooms. It is tempting to assume that a concern with a single discipline will in some way act as a unifying agent, but no two mathematics classrooms are exactly alike. Nevertheless, by focusing on culture, we can learn more about how the "invisible" components in the teaching and learning situation can contribute to or detract from the quality of the mathematical learning that takes place. An exploration of such issues as the influence of differing perceptions of mathematics as a subject, of teacher beliefs and actions, and of pupil perspectives may help clarify how some of these components contribute to the cultural context of the mathematics classroom.

The Anthropological Dimension

Culture is being considered here in its anthropological sense in that we are focusing on an aspect of the study of human culture; that is, we are concerned with what people do within a mathematics classroom. Scheffler (1976) draws attention to the organic metaphor of culture sometimes employed within an educational context, where education is largely interpreted in terms of growth. His suggestion is that culture in the anthropological sense may too easily be taken as similar to its use in the organic sense, where culture is interpreted in terms of renewal and adaptation, or of growth and sustaining equilibrium within an environment. The danger in using the metaphor in the organic sense in an educational context is that educational processes are essentially likened to "the processes by which individuals take on the environing culture" in order to ensure the continuity of that culture (Scheffler, 1976, p. 53). The importance in distinguishing between the two here is to draw attention to the fact that education *mediates* between the individual and his or her culture and that the invisible aspects of the cultural core are brought into play; as it were, by the teacher. The pupils being taught do not merely "take on" mathematics. In the context of the mathematics classroom, teachers act as agents of a particular culture, and in this role they make judgements and choices about aspects of that culture to which their pupils will be introduced—in this case, what mathematics will be taught, to whom, and how. These are the invisibles of the cultural core for which teachers have a responsibility.

Cooney, Goffree, Stephens, and Nickson (1985) have noted the importance of metaphors used in mathematics education and their potential effect on what we do. They suggest that such metaphors can reveal much about our basic beliefs, and by considering them, we can "increase our sensitivity to the possible roles researchers and teachers play" (p. 25). It is important in exploring the mathematics classroom from the perspective of the culture it generates to remember that we are concerned with the people in that setting and what they bring to it. We must increase our sensitivity to the importance of their hidden knowledge, beliefs, and values for mathematics education.

FROM CLOSED TO OPEN

As noted at the beginning of this chapter, subject specificity in studying the culture of a classroom is a relatively new phenomenon. Reiss (1978) states that, "Empirical research in education . . . is only hesitantly approaching the problem as a subject-matter specificity of teaching and learning processes. Research on the relationship between subject matter and patterns of social interaction in the classroom has been barely tentative, despite the fact that demands for it are repeatedly raised in related research resumés" (p. 400). Some years later, this situation is little changed and a plea for more classroom research related to the teaching and learning of mathematics is still being made (Good & Biddle, 1988). An encouraging example of an increase of interest in this area is Eisenhart's (1988) examination of the relationship between the ethnographic research tradition and research in mathematics education, which points to a greater concern with a study of mathematics classrooms and to a widening of the research base for such studies. Before exploring new directions in research methodology, the importance of a current change in beliefs and values about the nature of mathematical knowledge will be discussed because such beliefs and values related to mathematics as a subject are bound to affect much of what goes on in the classroom context.

Differing Views of Mathematics

One of the major shifts in thinking in relation to the teaching and learning of mathematics in recent years has been with respect to the adoption of differing views of the nature of mathematics as a discipline. Thom (1972) suggests that all mathematical pedagogy rests on a philosophy of mathematics, however poorly defined or articulated it might be. Even bearing in mind constraints imposed by being compelled to teach particular content, the way in which it is approached can be seen as a manifestation of a particular philosophy, a point implicitly acknowledged in the Non-statutory Guidelines (National Curriculum Council, 1989) of the recently developed National Curriculum in the United Kingdom. Some idea of the range of perceptions of the nature of mathematical knowledge can be gained by considering two traditions of mathematical thought - one of long-standing and one that has more recently come to the fore.

The "Formalist" Tradition. The view of mathematics that has informed and historically transfixed most mathematics curricula has been what Lakatos (1976) refers to as the formalist view, based on the epistemology of logical positivism. Hamlyn (1970) describes a central thesis of logical positivism that "all propositions *other than those about mathematics or logic* are verifiable by reference to experience" (p. 37). In other words, the foundations of mathematical knowledge are not seen to be in any sense social in origin, but lie outside human action in what Lakatos (1976) calls a "formalist heaven" (p. 2). Mathematics is considered to consist of immutable truths and unquestionable certainty. This in turn means that much of what we know as mathematics, such as problem solving, and what is commonly understood as mathematics in "the full spectrum of its relationships to science, to technology; to the humanities, and to human life" (Wittman, 1989, p. 298) is not recognised. Equally important is the fact that such a view does not take into account how mathematics changes and grows - it is waiting "out there" to be discovered. In short, mathematics as a way of knowing and interpreting our experience is discounted and becomes removed from human activity and the context of everyday life.

Hersh (1979) notes that "the criticism of formalism in the high school has been primarily on pedagogic grounds." He goes on to state that "all such arguments are inconclusive if they leave unquestioned the dogma that real mathematics is precisely formal derivations from stated axioms.... The issue then, is not, what is the best way to teach, but what is mathematics really all about" (p 18). Others have considered the implications of such a view (Bloor, 1976; Lerman, 1990; Nickson, 1981; Plunkett, 1981), some of which will be discussed below in relation to how this view affects the classroom context and the culture engendered.

A "Growth and Change" View of Mathematics. Lakatos' views have been responsible for much of the new direction in thought about the nature of mathematical knowledge in recent years. He is a proponent of Popper's (1972) view of objective knowledge, where knowledge is seen as resulting from competing theories that are proposed, made public, and tested against other theories and held to be true until "falsified" by a better theory. Kuhn (1970) goes beyond this by identifying the importance of the subjective aspect involved in the selection of theories in the first place. He emphasises the important part that judgement and choice have to play in such a situation and hence the role of different value systems. He suggests that "in many concrete situations, different values, though all constitutive of good reasons, dictate different conclusions, different choices" (p. 70). This `value' dimension of how knowledge comes into being, and also how it comes to be superseded and changed, means that it is not only a social phenomenon but a cultural one.

Mathematics educators in recent years have increasingly been drawn to consider mathematical knowledge from this persepctive. Researchers such as Confrey (1980), Wolfson (1981), Nickson (1981), Pimm (1982), Lerman (1986), and Cobb (1986) have been among those to study the implications of such a view for the mathematics classroom. Others such as Orton (1988) have explored the issue of how such theories can contribute to the study of fundamental issues in mathematics education. We shall now go on to consider how this view and a formalist view manifest themselves in the culture of the mathematics classroom.

Shared Views of Mathematics in the Classroom

Referring to the work connected with teaching cultures, Feiman-Nemser and Floden (1986) state that "teaching cultures are embodied in the work-related beliefs and knowledge teachers share - beliefs about appropriate ways of acting on the job and rewarding aspects of teaching, and knowledge that enables teachers to do their work" (p. 508). Whether or not teachers overtly identify a particular view of the nature of mathematics, they must hold beliefs and values with respect to the subject that influence how they teach it. These will affect what content they select, whether they consider it accessible to all pupils, and how they chose to make it accessible to them. It is reasonable to assume their actions throughout these processes will reflect their personal perceptions and beliefs related to the subject and its pedagogy. This has been confirmed to an extent by Thompson (1984), who investigated high school teachers' beliefs and their classroom teaching. She found that "the observed consistency between the teachers' professed conceptions of mathematics and the manner in which they typically presented the content strongly suggests that the teachers' views, beliefs, and preferences about mathematics do influence their instructional practice" (p. 125).

Evidence of Practice in the Mathematics Classroom. An analysis of classroom practice in the United Kingdom carried out over the past 15 years (Ashton, Kneen, Davies, & Holley, 1975; Department of Education and Science, 1978, 1979; Galton, Simon, & Croll, 1980; National Committee of Inquiry, 1982; Ward, 1979) suggests that the view of mathematics being projected was that of a linear subject, mainly concerned with mechanistically teaching facts and skills predominantly related to number, and generally characterised by paper-and-pencil activity, even where commercially published schemes were not in use. In the United States, Porter, Floden, Freeman, Schmidt, and Schwille (1988) have noted a reliance on mathematics textbooks in elementary schools and an associated emphasis placed by teachers on computational skills; time spent on teaching these skills takes up approximately 75% of class mathematics time. They note that "the lack of balance in teacher attention to conceptual understanding, skills, and applications is problematic and should be addressed" (p. 106).

Although this may be limited evidence upon which to make judgements about shared views of mathematics and the classroom cultures they generate, some inferences can be drawn. Whether or not overtly identified, it appears that the shared view of mathematics reflected in these particular classrooms is close to a formalist perspective insofar as it tends to concentrate on the formalities of the number system and the abstract manipulation of number.

The evidence does not suggest a classroom environment that reflects mathematics as "an integral part of human culture" (Wittman, 1989, p. 298). A heavy reliance on textbooks and teaching number facts is not conducive to teacher-pupil exchange or to interaction among pupils themselves; it is more conducive to an environment dominated by a concern with right or wrong answers. It could be argued that teachers in these classrooms adopt particular teaching strategies because of their supposed effectiveness, regardless of their personal teaching philosophy. Whether or not this is the case, the lack of success in achieving desirable results from the adoption of methodologies that reflect this formalist perspective has led to recent inquiries in the United States and the United Kingdom to study the failure of mathematics curricula in meeting the needs of these respective societies (National Committee of Excellence in Education, 1983; National Committee of Inquiry, 1982).

Where mathematics as a discipline has been perceived in formalist terms, it has on the whole remained inaccessible to teachers and hence to pupils. The traditional detachment of mathematical content from shared activity and experience so that it remains at an abstract and formal level, erects a barrier around the subject that removes it from other spheres of social behaviour. The messages conveyed in approaching the subject in this way are based upon assumptions that have been accepted unquestioningly and from which no deviation is permissible. The classroom culture that evolves will inevitably mirror this unquestioning acceptance. Brown and Cooney (1982) give an interesting example of this through an incident in which a geometry teacher was struggling with a below-average class. When it is suggested that the teacher downplay the importance of proof and engage in a more active approach to doing geometry, she resisted the idea because she had developed the strong belief that proof is necessary to learning geometry for all students, even though her students would not be tested on it. Brown and Cooney conclude that the intensity of the teacher's beliefs precluded considerations other than prooforiented ones. The visibility and acceptance of what is done and not done in mathematics, and of what is right and wrong, are potent factors in stopping teachers from engaging in activities that they may instinctively feel are appropriate but that might challenge the supposedly inviolable essence of mathematics as they themselves were taught it.

New Directions. In recent years following the Cockcroft Report (National Committee of Inquiry, 1982) in the United Kingdom and building on previous good practice, new methodologies have been developed and their adoption encouraged in mathematics curricula in an attempt to improve a situation where past success had been limited. Teachers have been encouraged to ensure that as much mathematics teaching as possible is based upon shared activity and that the subject is presented as open to discussion, investigation, and hypothesis (Department of Education and Science, 1985).

This is an approach that more closely reflects a growth-and-change view of the nature of mathematics, where mathematical knowledge (even at classroom level) is not held to be exempt from interpretations that require "reconsideration, revision and refinement" (Toulmin, 1972, p. 50). Teachers (especially at primary level) are coming to value the importance of discussion and problem solving and are encouraging children to generate theories and test them; indeed, teachers themselves generate theories (Alderson, 1988; McGrath, 1988). The decisions that are made in implementing this kind of curriculum, however, rest upon teachers' confidence in the appropriateness of doing so. It is clearly important that they should be able to give reasons for their choices and to identify criteria for judging the worthiness of what they do. The beliefs and values underlying this perspective have to be made more clearly identifiable to them to sustain their confidence in their professional judgement as teachers of mathematics. A point has been reached where a rationale is emerging that can offer them the theoretical support they need in sustaining that confidence. The fundamentals of such a rationale already exist, but need to be clearly articulated and made explicit and accessible.

Social Contexts. The implementation of curricula that reflect a growth-and-change perspective of mathematical knowledge will affect the social context of the mathematics classroom by implicitly encouraging the active participation of all concerned. If mathematics is characterised in terms of openness and of questioning and testing ideas, what is necessarily involved is a sharing of ideas and problems among pupils and between pupils and teacher. Although it is clear that specific concepts and skills have to be learned, the learning and application of these concepts and skills can become more purposeful because of this shared nature. The context becomes more purposeful rather than artificially contrived, and it gives rise to the possibility for increased discussion that is considered to be an important aspect of the mathematical learning process, particularly in the early years (Desforges & Cockburn, 1987). More will be said about teachers and pupils in the classroom later in this chapter, but the point being established here is how differing perceptions of the nature of mathematics (that is, the subject itself) can influence the culture of the mathematics classroom and why these perceptions must be taken into account in order to understand fully that culture.

Reiss (1978) notes that "in mathematical education, it is not only important what is known, but how it is known, and how this knowledge is reproduced" (p. 394). She suggests that in order to explore issues of this nature, what needs to be studied are the different forms and levels this interaction takes in classrooms across different subject areas. She notes research that suggests that "different behaviours are socially rewarded" in different subjects at each level (p. 405). These levels of interaction are characterised by degrees of negotiation between the participants about the rules for interpreting and evaluating actions in the classroom. The first level is where interaction occurs most frequently, and this takes place in humanities and social sciences classes. The second level is where participants move within an established and predetermined system of rules, and it is at this level that "social exchange in instructional situations in mathematics and natural sciences will predominantly take place" (Reiss, 1978, p. 405).

This analytical structure helps clarify the important effect specific subjects have on the interactive social processes within classrooms. The fact that levels of interaction within the mathematics classroom have been found to be constrained by an established structure of rules also suggests that this is a reflection of mathematics perceived in somewhat rigid and traditional formalist terms. However, with a shift in perceptions about the nature of the subject implicit in new methodologies and approaches within curricula, it is arguable that we are moving from the second level of social interaction in mathematics classrooms to the first. It is likely that when mathematics instruction is taken beyond establishing facts and practising skills to an approach using more openness, investigation, problem solving, and critical discussion, there will be more social interactions, more negotiation, and more emphasis upon shared interpretation and evaluation of what goes on in mathematics classrooms.

Summary

The differing views held by teachers and pupils in relation to the nature of mathematical knowledge are an important component in the culture of a mathematics classroom, since they are linked with the way mathematics is taught and received. One perspective may result in a classroom context which could be described as "asocial" insofar as it emphasises the abstractness of mathematics to be done individually and more or less in silence by pupils in the classroom. Another view emphasises the social aspect of the foundations of mathematical knowledge that precipitates a different kind of teaching and learning context in which the mathematics is sometimes done by individuals but can also be shared and open to question, challenged, discussed. explored, and tested. These examples may represent the two extremes, but they serve to indicate the potential power exerted by beliefs about mathematics on the teaching of mathematics

FROM TRANSMISSION TO PARTICIPATION

In recent years, educational research has become increasingly concerned with the study of classroom processes and the complexity of interactions within them. Lortie's (1975) sociological study of the schoolteacher, as well as other work such as that of Nash (1974) and Rutter, Maughan, Mortimore, & Ouston (1979), has helped to establish a background against which to consider issues of this nature. This increased interest in social and interpersonal aspects of the classroom has led to a shift in the nature of research undertaken, and it recognises the importance of social interactions within classrooms and of what pupils and teachers bring to them.

Shifting Research Perspectives

Kilpatrick (1988) offers a succinct description of three approaches to educational research that include the behaviourist tradition, the interpretivist view, and action research. Where the behaviourist tradition is concerned, he states that "the goal is to uncover law-like regularities in educational phenomena; the methods are aimed at specifying behaviour and analysing it into components. The world is a system of interacting variables whose variation can be controlled experimentally and modelled mathematically." (p. 98). Brown, Cooney, and Jones (1990) refer to a broader category of methodologies of this type as positivist.

In contrast to this, as Kilpatrick (1988) points out, the interpretivist view sets out to "capture and share the understanding that participants in an educational encounter have of what they are teaching and learning. "Action research, on the other hand, adopts the so-called critical approach" and seeks "not merely to understand the meanings participants bring to the educational process but to change those meanings that have been distorted by ideology." (p. 98) Cohen and Manion (1980) point out that is difficult to ascribe a "comprehensive definition" to action research "because usage varies with time, place and setting" (p. 174). Their interpretation is that action research is essentially small-scale and has to do with intervening in a real-world situation and closely examining the results of such intervention, but no specific link with ideology is made. Kilpatrick (1988) sums up the three approaches to research in the following way:

> The behaviorist stands apart from an educational encounter, aiming at general laws that will transcend time, place and circumstance. The interpretivist moves into that encounter, attempting to describe and explain it from a nonjudgmental stance. The action researcher enters the encounter with an eye toward obtaining greater freedom and autonomy for the participants. (p. 98)

He concludes that it would be unwise to view any of these approaches as existing in entirely separate compartments or as representing a shift away from “hard science." Trying to categorise research in mathematics education in line with these approaches would be extremely difficult and thus illustrates Kilpatrick's point. However, the application of new research techniques in the mathematics classroom has important consequences, in the long term, for the practicalities of how mathematics is taught and learned. Research is moving along a spectrum of activity from a position where the individual *acts* of teachers and pupils are studied, to a holistic approach where all teacher-pupil and pupil-pupil *interactions* are scrutinised together with the values, beliefs, and attitudes they bring to the situation. A brief consideration of each of the research paradigms helps to identify their potential to contribute to our knowledge of the culture of the mathematics classroom.

The Positivist Perspective in Mathematics Education Research. From the preceding interpretation of the behaviourist research paradigm, it can be seen that rather than focusing on the "whole" of a teaching and learning situation, research methodology of this type is concerned with how teachers and students function within the classroom. The purpose in offering the following examples related to the mathematics classroom is to identify the kinds of questions the methodology addresses. The intention is not to imply that researchers cited do not recognise the importance of social or cultural phenomena within the mathematics classroom, or that their research falls neatly into a single category. However, such studies do entail a choice with respect to topics for research that are seen potentially to be most fruitful. As such, they are offered as examples of research priorities and, at the same time, of the kinds of messages they may convey to teachers about what is important in mathematics education.

Many such studies have been concerned with exploring how students learn particular mathematical concepts such as conservation or operations with numbers (Bell, Fischbein, & Greer, 1984; Carpenter & Moser, 1983; Gelman, 1969) the stages of children's learning (Donaldson, 1978), the characteristics of individual learners (Trown & Leith, 1975), time spent on task (Peterson, Swing, Stark, & Waas, 1984), and seatwork (Anderson, 1981). The search has been for evidence of an optimum way to teach particular mathematics to particular children, with the aim to achieve transfer of these findings to the teaching of the same mathematics to similar pupils in similar situations. Other research of this nature has set out to identify learning styles, such as studies undertaken by Bennett (1976); McLeod, Carpenter, McCornack, and Skvarcius (1978), and those reported in Biggs (1967) and Shulman (1970). In general, by focusing on individuals, the search is for clusters of common characteristics from which to generalise about particular types of teachers or learners. Cooney et al. (1985) state that in employing such a paradigm for research, "importance is placed on the revelation of general truths and principles," and it is judged according to its power to "yield predictions and useful prescriptions" (p 25). Hence the value of research of this nature for the classroom teacher is seen to lie in its predictive nature and its potential for offering insights into successful ways of teaching mathematics.

Begle (1979) casts doubt on the relevance of much of this early work in the literature survey in which he set out to identify critical variables in mathematics education. He found that there was little correlation between the many teacher characteristics and variables identified and the effectiveness of teaching mathematics as measured by higher pupil achievement. He indicated that "the very concept of the effectiveness of a teacher may not be valid" (p. 37). This has not deterred further concern with the effectiveness of the individual teacher, however, as studies by Good, Grouws, and Ebmeier (1983), Berliner (1986), Leinhart (1988). and others have shown. Many studies of this kind have involved the observation and analysis of the behaviour of individual teachers who have been successful or "expert," with a view to identifying behaviours that may be adopted by others. Although the aim of all such studies is to produce useful insights into successful teaching practice, the message conveyed to teachers may be that there is some ideal way of teaching mathematical content. As Lampert (1985) puts it, teachers who do look to this kind of research for guidance may hold the view that "the teacher's work is to find out what researchers and policymakers say should be done with or to students and then to do it.... If the teacher does what she is told, students will learn" (p. 191).

Further exploration has focused on the effects of teachers' personal mathematical knowledge (Bassham, 1962; Malle, 1988), teachers' attitudes and expectations (Bishop & Nickson, 1983; Good & Brophy, 1978), and teachers' perceptions of and beliefs about mathematics (Brown & Cooney, 1982; Nickson, 1981; Thompson, 1984) upon what they do in the classroom. Some of this research has been observational in nature and has sometimes drawn upon work done in classrooms in general and been applied to the context of the mathematics classroom in particular.

From the point of view of the culture of the mathematics classroom, research on the actions of teachers and how these actions relate to pupil outcomes is likely to tell us little about the shared invisibles such as values, beliefs, and meanings that combine to create that culture. Clearly. if results of this type of study are made accessible to teachers, they may select from the conclusions to try to improve their classroom practice. However, what informs their judgement for making a selection in the first place - that is, the *reasons behind* particular choices - are important in understanding the classroom as a whole. It is these "work-related beliefs and knowledge teachers share," to which Feiman-Nemser and Floden (1986, p. 508) refer, that make a strong contribution to establishing the culture of a mathematics classroom.

A Constructivist Approach. The recent increased interest in constructivism as a theoretical perspective in mathematical pedagogy has contributed to the shift in the focus of research towards interactive processes and away from a behaviourist paradigm. Steffe and Killion (1986) state that, from the constructivist perspective, *"mathematics teaching* consists primarily of the *mathematical interactions* between a teacher and children" (p. 70). The teacher has an intended meaning to impart to the children, who interpret it and adjust it to their personal mathematics schemes, thereby constructing their own knowledge. This model of mathematical learning has formed the basis for a considerable body of research that includes work related to assessment (Brown, 1986) as well as to the learning of specific mathematical concepts (Herscovitz & Vinner, 1984; Von Glasersfeld, 1981). In spite of the interactive character of this perspective, the focus remains on the researcher or teacher, the learner, and specific mathematical concepts. and usually does not extend to the whole classroom context in which teaching and learning take place.

Desforges (1985) suggests that research concerned with matching pupils with tasks to achieve improved outcomes is based on prescriptive models of teaching derived by learning theories and "entirely ignore[s] the constraints on the teachers and the social conditions of the classroom" (p 96). Valuable detailed knowledge can be gained by studying how a pupil receives and accommodates some mathematical input, but the perceived relevance of such knowledge to teachers and the problem of its accessibility to them remains. The broad acceptance of children as constructors of their own mathematical knowledge has not been accompanied by a similar view of the teacher's research knowledge.

An Interpretivist Paradigm

It has been argued elsewhere (Nickson, 1988a) that bringing teachers into the arena of research activity can be an important step in increasing their understanding of research processes and results and their relation to classroom practise. It is the possibility of this type of collaboration that has been a factor in attracting researchers to the use of descriptive or naturalistic methodologies. Eisenhart's (1988) discussion of the ethnographic tradition of research in relation to mathematics education is particularly relevant here because it highlights the anthropological nature of the questions addressed in studying the culture of the mathematics classroom and it provides a valuable framework against which to discuss the interpretivist approaches to research increasingly being adopted in the field.

Ethnographic Research Methodology and Mathematics Education. Eisenhart (1988) makes the point that "few educational researchers have actually undertaken ethnographic research" (p. 99), which is defined as "the disciplined study of what the world is like for people who have learned to see, hear, speak, think, and act in ways that are different" (Spradley, 1980, p. 3). Spradley holds that ethnographic research has to do with learning *from* people, not from studying *them.*

The four methodologies used for data collection within this paradigm are identified as participant observation, ethnographic interviewing, a search for artefacts (available written or graphic materials related to the topic of study), and researcher introspection. These can be supplemented with others such as surveys and questionnaires. Data collection and analysis are carried out together throughout the study so that what is being tested is "an emergent theory of culture or social organisation." The ultimate goal of the research is "a theoretical explanation that encompasses all the data and thus provides a comprehensive picture of the complete meanings and social activity" (Eisenhart, 1988, p. 107).

Eisenhart acknowledges that researchers in mathematics education are identifying questions for which an ethnographic approach is appropriate, but suggests that the tendency for these researchers is "to use case studies, in-depth interviews, or in-classroom observations without doing what most educational anthropologists would call ethnographic research" as described above (p. 99). She goes on to show how, when "ethnography is placed within the context of interpretivism and cultural anthropology and then compared to traditional educational research and psychology," it becomes clear why ethnography as a methodology has been difficult to adapt to research in mathematics education. There is a long tradition of other well-established research paradigms in the education field with which it has to compete. Nevertheless, as the questions being addressed are seen to be increasingly appropriate for study using ethnographic methodology, stronger links will be forged between the fields of educational anthropology and mathematics education.

Interpretivist Studies. If we accept what Feiman-Nemser and Floden suggestion that the "complexity and immediacy of classroom events" (1986, p. 515-516) are what provide the link between teachers' beliefs and behaviour patterns, then it is these events that must be studied in order to gain an insight into the nature of the culture of a mathematics classroom. Eisenhart (1988) points out that the mixture of methodologies usually employed in interpretivist studies effectively limits them to a case study of a single class or of a small number of pupils or teachers. The inherent danger of the approach is the temptation to extrapolate from the local to the global (Davies, 1983). Davies describes some of the earlier work of this type, which he suggests has been detrimental to the field because researchers have jumped from "their little bits of the here-and-now to their big story" and in the process of so doing, have left "some of their actors (usually teachers) pretty trampled" (p. 133). If such approaches are to provide a way of bridging the gap between the mathematics teacher and the researcher, they must be used with sensitivity and care. They have the potential to involve teachers as partners in research, keeping in mind that it is the teacher's domain being studied and about which judgements will be made.

An example of the kind of public soul-searching teachers are often asked to engage in comes from Desforges and Cockburn's (1987) study of mathematics teaching in first schools in the United Kingdom. In this instance, the teacher sets out to establish the names of three-dimensional shapes. Following several acceptable student examples of sphere-shaped objects, Samantha suggests a circle. "Mrs G. thought, 'Help! This is going to be a lot more difficult than I thought. How am I going to explain the difference between a circle and a sphere?' " (p. 113). Later on, "head" is suggested as an example of a sphere. This time, Mrs G. says:

> At first I was pleased. That was a nice idea. It was quick to relate it to spheres. Then I thought that a head is not a perfect sphere . . . I did not want them to get the wrong ideas at the beginning - they are really capable of that. At the same time I did not want to bog them down with pernickety bits and pieces. It was hard to sort of weight that out. I was really struggling. (p 113-114)

This brief extract conveys both the immediacy of the dilemma that faced this teacher and her openness in admitting her dilemma. Similar dilemmas have been explored by Lampert (1985). This example shows that the teacher is trying to share what went on in her mind and to articulate her reason for actions in resolving the knowledge-practice conflict in which she finds herself. Lortie (1975) refers to "teachers' recurrent doubts about the value of their work with students" and states that in asking teachers to assess their own performance in his study, "no aspect of the teacher's work evoked as much emotion as this issue of assessing outcomes" (p 142). Clearly the intention of research is not to add to teachers' doubts and anxiety. This kind of situation, where the teacher finds herself justifying her choices, highlights the importance of sensitivity on the part of the researcher in interpretivist methodologies.

In Delamont 's (1981) view, familiarity is the greatest danger in descriptive methodologies. In discussing their wider use in educational settings, she suggests that one way to enhance the chances of obtaining meaningful information is to study "unusual, bizarre, or 'different' classrooms" (p 74). This approach contrasts with mathematics classroom studies with a more specific focus, for example, on the expert or the novice teacher (Berliner et al., 1988), where the focus is predetermined. In the latter kind of study the expert teachers have been singled out according to the degree of achievement of pupils in their charge, and the actions of these teachers are studied to search for commonalities that could account for their pedagogic success. If Delamont's (1981) ideas were implemented, the situation could be turned around, and a class might be identified as "unusual" because of the generally high level of mathematical achievement of the pupils. The question to be explored would be, "What is there about *this classroom* that produces these results?" The whole context, including the pupils' perspectives and what they bring to the situation in terms of beliefs and expectations, would be studied, not just the perspectives and actions of the teacher. By taking these phenomena into account and exploring the interactive processes within the class as a whole and at different levels, new meanings and insights can emerge that have particular relevance to the cultural aspects of the classroom. For example, it may be that the success of the class has as much to do with the fact that the teacher and pupils share a common perception of what mathematics is about as it has to do with specific actions on the part of the teacher.

A good example of a valuable insight gained within an interpretivist study is provided by Yates (1978) when she describes her intervention in a classroom situation in which Linda is exploring the making of cubes from six-square nets. Yates asks questions to find out how Linda is approaching the problem and realises that she has upset the girl's equilibrium. In her reflections on the situation, Yates asks herself why she intervened and concludes that it was because Linda's approach to the problem was not typical; she was approaching the problem differently than others. Yates identifies the uncertainty as her own and poses the question, "How often do we project our uncertainties onto our pupils?" (p. 62). This is a powerful question that many teachers of mathematics could benefit from asking themselves at frequent intervals. It is immediately accessible to the teacher, and the situation from which it was drawn is identifiable as one which teachers meet every day. Important phenomena such as the projection of our own uncertainties onto our pupils often become embedded in the exchanges and structures within the institutionalised classroom setting and then become lost as part of our pedagogical knowledge. They are an important part of the classroom processes that need to be brought to the surface. The self-examination to which Yates submits herself in this extract is a good example of "researcher introspection" that Eisenhart (1988) identifies as a component of the ethnographic tradition, and which in this particular situation clearly produced a valuable insight. The nature of teacher intervention conveys powerful messages to pupils. In this case, the pupil sensed disapproval on Yates' part that suggested unhappiness with the way she was proceeding. Yates was concerned that Linda would fail to achieve a satisfactory solution to the problem. Linda in turn was beginning to lose faith in her own problem-solving strategy, and her confidence in doing mathematics was undermined.

Good and Biddle (1988) suggest that "observational research has not been utilised sufficiently to improve mathematics education" and that it has the potential to "yield better theories for understanding the learning of mathematics and other subjects and can produce more adequate models for improving teaching" (p 114). The foregoing examples provide some indication of how observation can also lead to a more reflective approach on the part of teachers and researchers and can impart a deeper understanding of the processes that contribute to the culture of mathematics classrooms and theories about them.

Action Research

Two definitions of action research were offered earlier (Cohen & Manion, 1980; Kilpatrick, 1988). Kilpatrick identifies its current popularity in Australia and it has also appeared in curriculum development research in the United Kingdom over the past few decades (Shipman, 1974). Although action research does not appear to figure strongly in educational research in the United States at the moment, it has not been ignored. Cohen and Manion

(1980) state that "the scene for its appearance began to be set in [the United States] in the 1920s with the application of the scientific method to the study of educational problems, growing interest in group interaction and group processes, and the emerging progressive movement" (p. 176). It is still characterised by a concern with groups and their interaction.

The importance of action research in the context of the culture of the mathematics classroom is the potential it holds for the involvement of the teacher in the research process. Cohen and Manion (1980) suggest five broad categories of purpose for action research in schools and classrooms: (1) diagnosing and remedying problems in specific situations; (2) providing inservice training to sharpen analytical powers and self-awareness of teachers; (3) injecting innovative approaches to teaching and learning into an ongoing system; (4) improving poor communication between the practising teacher and the researcher and remedying the lack of clear prescriptions provided by traditional research; and (5) providing an alternative to more impressionistic and subjective approaches to solving problems in the classroom. It is clear that teachers can play a central part in addressing these kinds of problems in the classroom when they act as "both practitioner and researcher in one" (p 177). Action research differs from interpretivist research methodologies in that it relies both on observational and behavioural data and "is therefore empirical," so that over a period of time "information is collected, shared, discussed, recorded in some way, evaluated and acted upon" (p. 179-180). The overlap of methodologies emphasises the danger noted earlier of trying to be too specific in typifying research activities. Action research combines characteristics of both positivist and interpretivist paradigms.

The participant nature of action research with its direct involvement of teachers means that studies of this kind can be carried out by teachers researching in their own classrooms and schools, sometimes undertaking an inservice route towards a higher degree. An example of such a study with mathematics as a curriculum focus is an investigation of the effectiveness of parental involvement in a program of mathematics homework over a period of time and with specific teaching or learning objectives in mind (Alderson, 1988). Another teacher explores underachievers in mathematics in her own school (McGrath, 1988). These studies have involved the interaction of the teacher with other teachers, with pupils and parents, and with college researchers. The analytical and reflective skills and habits developed by teachers engaging directly in the research process have great potential for positively influencing their attitudes towards research generally. By developing selfconfidence in the field, teachers are likely to be ready to question and evaluate other research rather than to ignore it, or worse, to accept it unquestioningly.

Summary

As long as the main expectations of research in mathematics education are linked with its potential prescriptive power, the values or the essentially descriptive nature of other perspectives will remain questionable to many. This overview of approaches to research related to mathematics teachers shows how increased interest in the adoption of an interpretivist paradigm could prove fruitful in giving us insight into the culture of the mathematics classroom. Brown, Cooney, and Jones (1990) suggest that interpretivist research also has the potential to influence mathematics teacher education in several ways. It can provide a deeper understanding of meaning-making processes of teachers and thereby provide a basis for constructing teacher education programmes that are responsive to teachers' beliefs and needs. The hidden beliefs and meanings that all classroom participants bring to the teaching and learning situation must first of all be identified before they can be responded to. The interpretivist research paradigm seems particularly suited to the task.

FROM ACCEPTING TO QUESTIONING

Feiman-Nemser and Floden (1986) point out in their explorations of teaching cultures that it is important to be aware of norms "that govern social interactions between teachers and other role groups" (p. 508). The most immediate and significant "other role group" for teachers are their pupils, who exert the most important external constraint upon them" (Bishop & Nickson, 1983, p. 15). The developments described in the previous two sections suggest that pupils are increasingly playing a more active role in the mathematics classroom and potentially contribute much more overtly to the processes that combine to make up its culture. Judgements and choices are made by pupils as well as teachers.

Pupils' Perceptions of Mathematics: Accepting

Desforges and Cockburn (1987) provide an example of how the context in which mathematics is done conveys messages about the subject itself. As part of their study they probed children's perceptions of mathematics. The children were working from commercially published schemes, and the researchers found that in spite of the fact that teachers encouraged work at a steady pace, the children were "very keen to race on through the scheme's workbooks" (p. 101). The following is an extract from an interview with a pupil engaged in using such a workbook:

Interviewer: Why did you measure these things?
Kelly: For the book.
Interviewer: Why do you think it's in the book?
Kelly: It's our work. (p. 100-101)

Mathematics is seen by Kelly not as the practicality of measuring to gain any relevant or useful information or to solve a real problem; it is what the book "tells her to do" and it is part of her "work." The children interviewed could recall very few instances when they had seen adults engaged in measuring. Desforges and Cockburn (1987) conclude that "they never saw adults doing mathematics and in any event they were far more concerned with the reality of their own world which, mathematically speaking, was the self-imposed business of getting on through the scheme" (p. 102). The children had been found not to rate discussion sessions in mathematics very highly, and Desforges and Cockburn (1987) suggest that "perhaps the children saw them as distraction from the main agenda." They go on to say:

> The content of the work was dominated by the commercial schemes used, in part because the structure of the schemes looked like a race track to the children and in part because, forming the only overtly assessed part of the mathematics curriculum, they could be taken to be the whole of what was meant by mathematics. (p 102)

Carraher (1989) makes a similar point when she writes:

> Although the teacher may introduce a topic by making reference to everyday situations, the situations are usually stripped of their basic sense. The loss of meaning does not result from the pretend situation, from the role-playing of a situation which is not part of school life, but from the stripping off of central elements of the situation, which are viewed as extraneous to mathematics. (p. 639)

A further example offered by Cobb (1986) shows how learning mathematics in a particular context affects a pupil's beliefs about mathematics. He suggests that if a pupil asks a teacher for confirmation about the rightness or wrongness of a piece of work as soon as it is completed, this is evidence of the fact that the teacher is seen by the pupil as an authority. However, it is also evidence of the pupil's perceptions of the nature of mathematics, that is, that mathematics is a kind of knowledge that can always be assessed in terms of correctness or incorrectness. This is similar to the pupils in the science example offered by Lampert (1985), where they were confused by the possibility of two right answers, although in this case it probably had more to do with the way in which the knowledge was being assessed. Cobb (1986) states that "this inference would be further substantiated if the teacher responds by attempting to initiate a Socratic dialogue and the student shows irritation or frustration" (p. 4) Mathematics is not seen to be a subject for exploration or discussion where the possibility of negotiation about methodology or content may exist. The expectation is that alternatives do not exist.

The first of the preceding examples shows how children's views are in the process of being shaped by the way mathematics is being presented, while the second is an illustration of the long-term effects of this kind of approach, where expectations become narrowly focused on the criterion of the "right or wrong" nature of mathematical knowledge. Cobb (1987) also reports a study related to classroom contexts in which arithmetic is being taught by teachers who have declared their expectations that the children will produce correct answers, and, in most cases, the teachers have prescribed the methods they expect students to use. Cobb refers to the fact that the children were faced with problems, but says that "these problems were not purely arithmetical in origin. They derived from the children's social interactions in classrooms characterised by imposition," and further suggests that "the children had the responsibility of adjusting their goals and activities to fit their teachers' relatively rigid expectations" (Cobb, 1987, p. 121). The view of mathematics being developed is rigidly determined by the specific requirements of the teacher, and rather than the context being one of activity, negotiation, and discussion, it is one of repression, where the use of the pupils' own strategies and potential for solving problems is discouraged. There is no room for manoeuvre, and should the pupil instinctively wish to approach a problem differently the teacher's clear expectations produce in the student an immediate internal conflict. This is reminiscent of Yates' (1978) dilemma identified earlier, when she realised that her interference with her pupil's problem-solving strategy upset the pupil's equilibrium.

Pupils' Perceptions of Mathematics: Questioning

Jaworski (1989) has studied mathematics classrooms in which an investigative approach has been adopted, which she links with a constructivist view of mathematics teaching and learning. In one case study she describes a lesson in which a class of 14- and 15-year-olds were exploring shapes that had the same area and perimeter. The class worked in groups of their own choice after negotiating with each other about which shapes they wished to study. The teacher, over a period of time with the class, had established the importance of looking for pattern in their mathematical work, which symbolised his "belief in mathematics being about expressing generality" (Jaworski, 1989, p. 151). In her observations of what was going on in the classroom Jaworski (1989) notes that "implicit in the interchange here was an emphasis on conjecturing, on trying out special cases and on seeking for generality" (p. 151). Two boys remained after the lesson to discuss some aspects of it, and the following comment about the teaching methodology was made by one of them to the researcher with the teacher present:

> To tell you the truth . . I mean, Mr xxxx's . . . a different kind of teacher completely. Before, you've had sums that you've been set . . . At first, to tell you the truth, I didn't like him as a teacher. I thought, 'No. Pathetic! - you know, this isn't maths - what's this got to do with maths? And as I've come along, I've realised that it's got a lot to do with maths. To have to learn rather than just have to sit and (say) 'Oh, I've done

50 sums today', 'I've done a hundred.' You don't bother about that now, just concentrate, and at the end of a lesson you've learned something . . . I've really progressed. (p. 153-154)

This extract indicates how beliefs about mathematics that the members of this class share with their teacher reflect a classroom context that differs considerably from those projected by pupils in the examples given earlier. The interactions between pupils and teacher and among pupils are based on a perception of mathematics that does not exclude problems that can be shared. However, what the pupil has to say also hints at the difficulties faced by teachers who hold constructivist views of mathematics and try to teach mathematics to pupils who have developed quite different perceptions from years of prescriptive teaching. If mathematics is seen by pupils to be solely about accuracy in doing number work or any other kind of mathematical work, it is bound to be difficult for teachers to alter such views. Pupils' expectations built up over the years can provide a considerable block to effecting such change. It is clear that such a conflict of views would manifest itself in some way in the culture of the classroom in which it exists. If teachers become aware of the potential for such conflict and of what the associated symptoms might be, they are then in a positiom to act upon them where they arise.

Pirie (1988) reports research that focuses on discussion as an aid to mathematical understanding. She describes how the teacher starts a lesson by promoting a whole class discussion by putting forward a mathematical idea or problem and keeping the discussion open so that "pupils are encouraged to offer their thinking to the rest of the class." The teacher remains as unobtrusive as possible and deflects questions back to the pupils. The intention is "that the pupils shall form their own intuitions about the structures of mathematics" (p. 2). The pupils then form small groups to pursue a solution to the problem. Pirie (1988) states that "the bedrock on which this style is built is the belief that knowledge must be actively constructed and not passively received if the pupils are truly to learn mathematics" (p. 2). She describes how one pair of girls share their discovery of how to divide fractions; she also discusses the strategies they evolve to cope with that problem. She suggests that concepts are understood not just by experiencing them, but by interpreting those experiences to each other and also, in this case, to the researcher.

In a classroom such as this, it is made clear to the pupils through the teaching practice used that their views and conjectures are as valuable as anyone else's. They are developing their perceptions of mathematics from those of their teachers and peers, and they are learning to articulate their own ideas and beliefs about them. They are actively engaged in doing mathematics and in the process, they are receiving messages that mathematics is about questioning, conjecturing, and trial and error. The tools necessary for engaging in these processes are learned as pupils advance and are in a context that is meaningful either because it is a problem that they have made their own through exploration and discussion, or because it is a real-life problem to be solved.

Summary

The studies referred to in this section show how the pupil's role in the mathematics classroom can vary dramatically according to the view of mathematics projected by the teacher. The linearity and formality associated with most teaching of mathematics from published schemes or textbooks tend to produce a passive acceptance of mathematics in the abstract, with little connection being made by pupils between their work and real life. The visibility of mathematics in terms of a "right or wrong" nature is accepted by pupils, and their main concerns seem to be with the quantity of mathematics done and its correctness. The culture of such a classroom seems to be fairly well defined, since the pupils' expectations are based on their acceptance of what appear to be clear messages from the scheme or the book and the teacher. The nature of the social interaction within the classroom is dominated by these perceptions.

Where beliefs about mathematics differ and where views of mathematics as socially constructed knowledge prevails, pupils take on quite a different role. The messages they receive are that they are expected to contribute their own ideas, to try their own solutions, and even to challenge the teacher. In such situations the culture of the mathematics classroom is likely to be more variable, since there exists the possibility for greater divergence of views. There is also greater potential for conflict in terms of what is agreed and shared, since the development of these views of mathematics is fairly recent and because not all pupils will necessarily accept them in the light of their previous educational experience in mathematics.

CONCLUSIONS

What has emerged in this study is that the culture of the mathematics classroom will vary according to the actors within it. The unique culture of each mathematics classroom is the product of what the teacher and pupils bring to it in terms of knowledge, beliefs. and values, and how these affect the social interactions within that context. It is all too easy to assume that these invisibles of the cultural core are shared by all the participants and that there is a harmony of views about the goals being pursued and the values related to them. The research suggests that there may once have been a fair degree of predictability about what could be expected, as well as uniformity in the views held and the methodologies they gave rise to. This resulted from a dominant perception of the nature of mathematical knowledge and the way it affected pedagogy, which in turn affected the social interaction within the classroom and, ultimately, the culture of that classroom.

However, the concurrent changes over the three broad fronts considered here mean that an increased potential for variation now exists within the context of the mathematics classroom with respect to what is taught, as well as how and to whom it is taught. There is more possibility for choice and more possibility that those choices will be guided by different beliefs and values. Consequently there will be greater variation in the cultures of mathematics classrooms. With this increased breadth of possibilities, however, comes increased potential for lack of consensus, in particular for mismatches between teachers' views and goals and those of pupils. This can result in what we might call productive (in terms of mathematical learning) classroom cultures and non-productive classroom cultures. If we are to gain anything from the study of the culture of the mathematics classrooms, it will come from an understanding of the factors that contribute to their productivity in this sense.

What Needs to be Done? Several issues have emerged in this study that can be explored in order to help to achieve this understanding. They are closely linked with the changes that have been discussed and that need to be addressed because of their potential for militating against a productive classroom culture and hence their potential for obstructing the effective teaching and learning of mathematics.

A New Rationale. There is a need for a clearly articulated rationale to provide a theoretical framework for new methodologies and approaches that are developing concurrently with changing views about the nature of mathematical knowledge. This rationale should be accessible to teachers in order to avoid the conflict of values that can arise when they are asked to adopt these new methodologies without understanding their foundations.

Potential Areas of Conflict It is important to be aware of the potential for conflicting views and their effects within the class room where, for example, a pupil may be expected to assume a changed role within the context of the mathematics lesson. Teachers' expectations differ depending on their beliefs and values, yet pupils are expected to adjust and accept them as valid because the teacher is seen as an authority. We need to develop strategies for coping with such situations and to consider ways of negotiating shared meanings and goals.

Messages Conveyed by Research. An increase in the adoption of new research methods that study the reality of the class room is needed. These methods can provide a rich opportunity for teachers to become drawn into research as equal partners and can allow them to explore questions about mathematics in their classrooms that they see as important. Topics for study in the past have conveyed messages that have placed a premium on the mechanics of maximising mathematical learning. With teachers as partners in research, the message conveyed will be that their ideas and expertise have credence and value. Through their participation, teachers will come to appreciate the full effect that their personal beliefs and values have on their mathematics classroom.

Repression of Pupils' Ideas in the Mathematics Class room. We need to be more aware of the subtle effects of social interaction in the classroom and how actions and comments by teachers can repress the pupil's individual mathematical thinking. New methodologies allow teaching and learning situations where the pupil is encouraged to challenge and question the teacher as well as other pupils. Implicit in this situation is the need to acknowledge and value what the pupil offers.

Accessibility. It has been clear throughout this chapter that matters related to the culture of the mathematics classroom all have to do, one way or another with the accessibility of mathematics both to teachers and pupils. The assumption is that all members of a given society have access to mathematical knowledge at some level. It is important to remember how vital that accessibility is to individuals and how the extent to which mathematical knowledge can add to the quality of their lives by helping to inform the judgements and decisions they make.

The importance of the culture of the mathematics classroom is implicit in the statement by Reiss (1978) that 'individual acquisition of mathematical concepts and skills in school will always be accompanied by an acquisition of orientations about the subject matter related and the social significance of the things learned" (p. 394-395). She is hinting at the hidden messages that form a main part of the culture of the mathematics classroom. Popkewitz (1988) writes in a similar vein when he states that "school mathematics involves not only acquiring content; it involves participating in a social world that contains standards of reason, rules of practice and conceptions of knowledge" (p. 221). As mathematics educators, we need to be more aware of the hidden social messages in what we do and the power of their influence on the young people we teach.

REFERENCES

Alderson, J.: 1988, *Parental involvement in children's mathematics* Unpublished master's thesis, Essex Institute of Higher Education, Brentwood, U.K.

Anderson, L. M.: 1981, *Student responses to seatwork: Implications for the study of students' cognitive processes* (Research Series No. 102). East Lansing, MI: Institute for Research on Teaching, Michigan State University.

Ashton, P., Kneen, P., Davies, F. and Holley, B.: 1975, *The aims of primary education: A study of teachers' opinions.* London: Macmillan Education Ltd.

Bassham, H.: 1962, Teacher understanding and pupil efficiency in mathematics - a study of relationship. *Arithmetic Teacher* 9, 383-387.

Begle, E. G.: 1979, *Critical variables in mathematics education.* Washington, DC: Mathematics Association of America and the National Council of Teachers of Mathematics.

Bell, A., Fischbein, E., and Greer, B.: 1984, Choice of operation in arithmetic problems: The effects of number size, problem structure and context *Educational Studies in Mathematics.* 15, 129-147.

Bennett, N.: 1976, *Teaching styles and pupil progress* London:Open Books.

Berliner, D. C.: 1986 In pursuit of the expert pedagogue. *Educational Researcher* 15. 5-13.

Berliner, D. C., Stein, P., Sabers, D., Clarridge, P., Cushing, K., and Pinnegar, S.: 1988, Implications of research on pedagogical expertise and experience for mathematics teaching. In D. A. Grouws, T. Cooney, and D. Jones, (Eds.), *Effective perspectives on research on teaching* (pp. 67-95). Reston, VA: National Council of Teachers of Mathematics.

Biggs, J. B.: 1967, *Mathematics and conditions of learning* Slough, U.K.: National Foundation for Educational Research (NFER).

Bishop, A. J. and Nickson, M.: 1983, *The social context of mathematics education: A review of research in mathematical education* (Part B), Windsor, U.K.: NFER-Nelson.

Bloor, D.: 1976, *Knowledge and Social Imagery* London: Routledge & Kegan Paul.

Brown, C. A., and Cooney, T. J.: 1982, Research on teacher education: A philosophical orientation, *Journal of Research and Development in Education,* 15 (4) 13-18.

Brown, M.: 1986, Developing a model to describe the mathematical progress of secondary school students (11-16 years): Findings of the graded assessment in mathematics project. In *Proceedings of the Tenth International Conference for the Psychology of Mathematics Education* (pp. 135-140). London, U.K.

Brown, S. I., Cooney, T. J., and Jones, D.: 1990, Mathematics teacher education. In W R Houston (Ed.), *Handbook of Research on Teacher Education* (pp. 639-656). New York:Macmillan.

Carpenter, T. P., and Moser, J. M.: 1983, The acquisition of addition and subtraction: A cognitive perspective. In R. Lesh & M. Landau (Eds.), *Acquisition of mathematics concepts and processes* (pp. 7-44), New York: Academic Press.

Carraher, T. N.: 1989, Negotiating the results of mathematical computations. *International Journal of Educational Research,* 13(6), 637-646.

Cobb, P.: 1986, Contexts, goals, beliefs and learning mathematics *For the Learning of Mathematics,* 6(2), 2-9.

Cobb, P.: 1987, An investigation of young children's academic arithmetic contexts *Educational Studies in Mathematics,* 18, 109-124.

Cohen, L., & Manion, L.: 1980, *Research methods in education,* London:Croom Helm.

Confrey, J.: 1980, *Conceptual change analysis: Implications for mathematics and curriculum inquiry* East Lansing, MI: Institute for Research on Teaching, Science-Mathematics Teaching Center, Michigan State University.

Cooney, T. J., Goffree, F., Stephens, M., and Nickson, M.: 1985, The professional life of teachers *For the Learning of Mathematics,* 5(2), 24-30.

D'Ambrosio, U.: 1985, Ethnomathematics and its place in the history and pedagogy for mathematics *For the Learning of Mathematics,* 5(1), 44-48.

Davies, W. B.: 1983, The sociology of education In P. Hirst (Ed.), *Eductional theory and its foundation disciplines* (pp. 100-137), London: Routledge & Kegan Paul.

Delamont, S.: 1981, All too familiar? A decade of classroom research *Educational Analysis,* 3(1), 47-68.

Department of Education and Science: 1978, *Primary Education in England A survey by H. M. Inspectors of Schools* London: Her Majesty' s Stationery Office.

Department of Education and Science: 1979, *Aspects of secondary education in England* London: Her Majesty's Stationery Office.

Department of Education and Science: 1985, *GCSE: The national criteria for mathematics* London: Her Majesty's Stationary Office.

Desforges, C.: 1985, Matching tasks for children. In S. N. Bennett and C. Desforges, (Eds.) *Recent advances in classroom research* (pp 92-104) London:Constable.

Desforges, C., and Cockburn, A.: 1987, *Understanding the mathematics teacher: A study of practice in first schools.* London:The Falmer Press.

Donaldson, M. C.: 1978, *Children's minds.* London:Croom Helm.

Eisenhart, M. A.: 1988, The ethnographic research tradition and mathematics education research *Journal for Research in Mathematics Education,* 19(2), 99-114.

Fasheh, M.: 1982, Mathematics. culture and authority *For the Learning of Mathematics,* 3(2), 2-8.

Feiman-Nemser, S., and Floden, R. E.: 1986, The cultures of teaching. In M. C. Wittrock (Ed.), *Handbook of research in teaching* (3rd edition, pp. 505-526). London: Collier-Macmillan.

Galton, M., Simon, B., and Croll, P.: 1980, *Inside the primary classroom* London:Routledge & Kegan Paul.

Gelman, R.: 1969, Conservation acquisition: A problem of learning to attend to relevant attributes, *Journal of Experimental Child Psychology* , 7, 167-187.

Good, T., and Biddle, B. J.: 1988, Research and improvement of mathematics instruction: The need for observational resources. In D. A. Grouws, T. J. Cooney, & D. Jones (Eds.), *Perspectives on research on effective mathematics teaching* (Vol. I, pp. 114-142) Reston, VA: National Council of Teachers of Mathematics.

Good, T., and Brophy, J.: 1978, *Looking in classrooms* New York: Harper & Row.

Good, T., Grouws, D., and Ebmeier, H.: 1983, *Active mathematics teaching* New York:Longman.

Hamlyn, D. W.: 1970, *The theory of knowledge* London:Macmillan Press.

Herscovitz, R., and Vinner, S.: 1984, Children's concepts in elementary geometry - a reflection of teachers' concepts. In B. Southwell et al. (Eds.), *Proceedings of the Eighth International Conference for the Psychology of Mathematics Education* (pp. 28-34) Darlinghurst, Australia.

Hersh, R.: 1979, Some proposals for revising the philosophy of mathematics *Advances in Mathematics,* 31(1), 31-50.

Jaworski, B.: 1989, *To inculcate versus to elicit knowledge* Actes de la 13e conference internationale, Psychology of Mathematics Education, Paris.

Joseph, G.: 1990, April, *The politics of anti-racist mathematics.* Paper presented at a conference on the Political Dimensions of Mathematics Education, Institute of Education, University of London, U.K.

Kilpatrick, J.: 1988, Editorial, *Journal for Research in Mathematics Education,* 19(2), 98.

Kuhn, T.: 1970, *The structure of scientific revolutions (2nd ed.).* Chicago:The University of Chicago Press.

Lakatos, I.: 1976, *Proofs and refutations.* Cambridge, U.K.: Cambridge University Press.

Lampert, M.: 1985 How do teachers manage to teach? *Harvard Educational Review* 55(2), 178-194.

Leinhart, G.: 1988, Expertise in instructional lessons: An example from fractions In D. A. Grouws, T. J. Cooney, & D. Jones (Eds.), *Perspectives on Research on Effective Mathematics Teaching* (pp. 47-66). Reston, VA: National Council of Teachers of Mathematics; Hillsdale, NJ: Lawrence Erlbaum.

Lerman, S.: 1986, *Alternative views of the nature of mathematics and their possible influence on the teaching of mathematics.* Unpublished doctoral dissertation, Centre for Educational Studies, King's College, University of London, U.K.

Lerman, S.: 1990, Alternative perspectives of the nature of mathematics and their influence on the teaching of mathematics, *British Educational Research Journal,* 16(1), 15- 61.

Levitas, M.: 1974, *Marxist perspectives in the sociology of education* London: Routledge & Kegan Paul.

Lortie, D. C.: 1975 *Schoolteacher: A sociological study* Chicago:The University of Chicago Press.

Malle, G.: 1988 *The question of meaning in teacher education.* Paper presented at the Sixth International Congress on Mathematical Education, Budapest, Hungary.

McGrath, J. E.: 1988, *Looking for a common denominator: A study of underachievement in the primary mathematics classroom.* Unpublished master's thesis, Essex Institute of Higher Education, Brentwood, U.K.

McLeod, D. B., Carpenter, T. P., McCornack, R. L., and Skvarcius, R.: 1978, Cognitive style and mathematics learning: The interaction of field independence and instructional treatment in numeration systems *Journal for Research in Mathematical Education,* 9, 163-174.

Mellin-Olsen, S.: 1987, *The politics of mathematics education.* Dordrecht, Holland: D. Reidel..

Nash, R.: 1974, Pupils' expectations of their teachers *Research in Education,* 47-61.

National Committee of Excellence in Education: 1983, *A nation at risk: The imperative for educational reform.* Washington, DC:National Institute of Education.

National Committee of Inquiry into the Teaching of Mathematics in Schools: 1982, *Mathematics counts (The Cockcroft report).* London:Her Majesty's Stationery Office.

National Curriculum Council: 1989, *Mathematics in the National Curriculum* London:Her Majesty's Stationary Office.

Nickson, M.: 1981, *Social foundations of the mathematics curriculum: A rationale for change.* Unpublished doctoral dissertation, Institute of Education, University of London, U.K.

Nickson, M.: 1988a, Pervasive themes and some departure points for research into effective mathematics teaching. In D. A. Grouws, T. J. Cooney, and D. Jones (Eds.), *Perspectives on research on effective mathematics teaching* Vol. 1, pp. 245-252, Reston, VA: National Council of Teachers of Mathematics.

Nickson, M.: 1988b, What is multicultural mathematics? In P. Ernest (Ed.), *Mathematics teaching: The state of the art* pp. 236-241, Lewes, Sussex:Falmer Press.

Noss, R.: 1988, The computer as a cultural influence in mathematical learning *Educational Studies in Mathematics,* 19, 251-268.

Orton, R. E.: 1988, Two theories of 'theory' in mathematics education: Using Kuhn and Lakatos to examine four foundational issues *For the Learning of Mathematics,* 8(1), 36-43.

Peterson, P. L., Swing, S. R., Stark, K. D., and Waas, G. A: 1984, Students' cognition and time on task during mathematics instruction *American Educational Research Journal,* 21(3), 487-515.

Pimm, D.: 1982, Why the history and philosophy of mathematics should not be rated X. *For the Learning of Mathematics,* 3(1), 12-15.

Pimm, D.: 1990 April, *Mathematical versus political awareness: Some political dangers inherent in the teaching of mathematics* Paper presented at a conference on the Political Dimensions Mathematics Education, Institute of Education, University London, U.K.

Pirie, S. E. B.: 1988, Understanding: Instrumental, relational, intuition, constructed, formalised. . .? How can we know? *For the Learning of Mathematics* 8(3), 2-6.

Plunkett, S.: 1981, Fundamental questions for teachers *For the Learning of Mathematics,* 2(2), 46-48.

Popkewitz, T. S.: 1988, Institutional Issues in the study of school mathematics: Curriculum research. In A. J. Bishop (Ed.), *Mathematics education and culture* pp. 221-249, Dordrecht, Holland: Kluwer Academic Publishers.

Popper, K.: 1972, *Objective knowledge - an evolutionary approach* Oxford:Oxford University Press.

Porter, A., Floden, R., Freeman, D., Schmidt, W., and Schwille, J.: 1988 Content determinants in elementary school mathematics. In D. Grouws, T. J., Cooney and D. Jones (Eds.), *Perspectives on research on effective mathematics teaching* Vol. 1, pp. 96-114, Reston, VA: National Council of Teachers of Mathematics.

Reiss, V.: 1978, Socialization phenomena in the mathematics classroom: Their significance for interdisciplinary teaching. In *Cooperation between science teachers and mathematics teachers* pp. 392-417, IDM Bielefeld Materialen und Studien Ban 16 University of Bielefeld.

Rutter, M., Maughan, B., Mortimore, P., and Ouston, J.: 1979, *Fifteen thousand hours: secondary schools and their effects on children* London:Open Books.

Scheffler, I.: 1976, *The language of education* Springfield. IL: Charles C. Thomas.

Shipman, M. D.: 1974, *Inside a curriculum project* London: Methuen.

Shulman, L. S.: 1970, Psychology and mathematics education In E. Begle (Ed.), *Mathematics Education* pp. 23-71, 69th Yearbook of the National Society for the Study of Education, Chicago: University of Chicago Press.

Smith, B. O., Stanley W. O., and Shores, J. H.: 1971, Cultural roots of the curriculum. In R. Hooper (Ed.), *The curriculum: Context, design and development* pp. 16-19, Edinburgh: Oliver & Boyd.

Spradley, J.: 1980, *Participant observation.* New York: Holt, Rinehart & Winston

Steffe, L. P., and Killion, K.: 1986, Mathematics teaching: A specification in a constructionist frame of reference. In L. Burton and C. Hoyles (Eds.), *Proceedings of the Tenth International Conference for the Psychology of Mathematics Education,* pp. 207-216, London: University of London, Institute of Education, U.K.

Thom, R.: 1972, Modern mathematics: Does it really exist? In A. G. Howson (Ed.), *Developments in mathematical education* pp. 194-209, Cambridge, U.K.: Cambridge University Press.

Thompson, A. G.: 1984, The relationship of teachers' conceptions of mathematics and mathematics teaching to instructional practice. *Educational Studies in Mathematics,* 15(2), 105-127.

Toulmin, S.: 1972, *Human understanding* Oxford: Clarendon Press.

Trown, E. A., and Leith, G. O. M.: 1975, Decision rules for teaching strategies in primary schools: Personality - treatment interactions. *British Journal of Educational Psychology,* 45, 130-140.

von Glasersfeld, E.: 1981, An attentional model for the construction of units and number *Journal for Research in Mathematical Education,* 12(2), 83-94.

Ward, M.: 1979, *Mathematics and the l0-year-old* Schools Council Working Paper No. 61. London: Evans-Methuen Educational.

Wittman, E. C.: 1989, The mathematical training of teachers from the point of view of education *Journal fur Mathematik—Didaktik,* 10(4), 291-308.

Wolfson, P.: 1981, Philosophy enters the classroom *For the Learning of Mathematics,* 2(1), 22-26.

Yates, J.: 1978, *Four mathematical classrooms: An inquiry into teaching method.* Southampton, UK.: Faculty of Mathematical Studies, University of Southampton.

Zaslavsky, C.: 1989, Integrating mathematics with the study of cultural traditions, In C. Keitel, P. Damerow, A. J. Bishop, and P. Gerdes (Eds.), *Mathematics, education and society* pp. 14-15, Paris: United Nations Educational, Scientific and Cultural Organisation (UNESCO).

Marilyn Nickson,
University of Cambridge Local
Examinations Syndicate,
Research and Evaluation Division,
Lloyds House,
95-96 Regent Street,
Cambridge CB2 1AW U.K.

R. NOSS

SETS, LIES AND STEREOTYPES

In an earlier version of this paper (Noss, 1989), I sketched the rationale for recent changes in the UK curriculum, reviewed its antecedents, and offered a critical reading of the response to such changes from mathematical educators. In Noss (1990), I developed this theme, and tried to show how the specificities of the new National Curriculum played a role in establishing the general priorities of those in a dominant position in UK society. In this paper, I attempt to draw broader and more general conclusions, and to discuss the ways in which mathematics curricula can be conceptualised from a political point of view.

A (TRUE) PARABLE OF OUR TIMES

I open the door of the bank and advance to the red and grey vinyl-covered desk marked *ENQUIRIES.* I smile at the red and grey-uniformed clerk, whose badge neatly proclaims her first name and she smiles at the computer screen which is neatly angled so as not to quite separate us. I start to enquire, but it seems that nothing can be done until the computer has been satisfied: I hand over my passbook and the 13 alphanumeric characters are expertly entered. Now Chris knows my name, where I live, how much money I have, how much I owe; now she will talk to me.

I start on my series of requests. 'Bear with me', Chris says as I finish the first question concerning my account. She walks over to a desk marked *ACCOUNTS* where another uniformed clerk is on the phone. Chris waits until the call is finished, and asks her something I cannot hear. When she returns both I and the computer are happy with the answer. I ask a second question. This, it turns out, necessitates a call to one of the helplines, whose codes are preprogrammed onto the multi-buttoned telephone in front of her. 'Bear with me' she says. After a short call, Chris knows the answer, tells me, smiles and punches some keys on the computer. My third and fourth enquiries generate a second call (to a different helpline) and a visit to the supervisor, apparently because my cheque is above the specified amount which Chris is allowed to handle unsupervised. The supervisor — whose uniform is several shades darker than Chris's, and who sports both her first and family names on her badge — looks at me, at my cheque, then back to me and nods. Chris places the passbook on a printer which is attached to the computer, and the machine punches information onto the book (by some miracle, it knows where to start typing on the page).

S. Lerman (ed.), Cultural Perspectives on the Mathematics Classroom, 37–49.

I am sent to queue for a teller, each of whom sits in a line behind a glass panel with their own computer. Their uniforms are like Chris's, but unlike her, the tellers are sedentary — they are tied to their machines. They chat to each other, crack jokes while their printers print, help each other with troublesome computers.
'The computer won't let me do that' my teller says, smiling.
Why not?' I ask.
'Bear with me' she says, and she calls the supervisor who comes to talk to me through the glass.
She explains: 'We can't do that' she says.
'Why not?' I ask.
'It's the computer' she explains again, 'it won't let us .. look'. She types the numbers on the keyboard: they disappear as she types *enter*. She is right: the computer won't let her. The computer doesn't explain why. The supervisor doesn't ask. Neither does Chris.

THE SOCIAL FUNCTION OF SCHOOLING

To its honour, Parliament has lately decided that England shall in future be governed by popular government...now that we have given them the power we must not wait any longer to give them the education. (Forster 1870, quoted in Maclure 1965, p.105).

Those who are not schooled to enjoy the blessing of a good society can only dispoil its institutions and corrupt themselves (Adler 1982, quoted in Popkewitz and Pitman 1986, p. 20).

These two quotations, separated by more than a century, provide a starting point for understanding the social function of schooling in Western societies. There has been no more hotly debated issue in the field of the sociology and politics of education than this question of the relationship between education and society. In this paper I can do no more than allude to the main strands of this debate, and to assert my own position.

In my view, one of the seminal insights into this issue has been provided by Harry Braverman's description of schooling and its relationship to the increasingly deskilled labour market into which its 'products' — the children — are inserted. His basic argument retains its force:

The schools as caretakers of children and young people are indispensible for family functioning, community stability, and social order in general (although they fulfill even these functions badly). In a word, there is no longer any place for the young in this society other than school. Serving to fill a vacuum, schools have themselves become that vacuum, increasingly emptied of content and reduced to little more than their own form. Just as in the labour process, where the more there is to know the less the worker need know, in the school the mass of future workers attend the more there is to learn, the less reason there is for teachers to teach and students to learn. In this more than any other factor — the purposelessness, futility, and empty forms of the educational system — we have the source of the growing antagonism between the young and the schools which threatens to tear our schools apart. (Braverman 1974, p.440)

Braverman's chilling and apocalyptic account of schooling hinges crucially on his notion of the 'deskilling' of the labour process. He argues that the twentieth century has been characterised by rapid changes in technology which has created all kinds of new jobs and skills, and new layers of people — the 'new middle class' — to work within them. The imperatives of the system demand that ever more complete control of the labour process is gained by management at the expense of the exercise of craft knowledge or independence on the part of the worker (including the new middle class itself). The production process has become degraded; less skill and training is required; the cost of labour is cheapened; more alienation from the work process has resulted.

Braverman's central argument is that there has been an increasing polarisation between highly skilled and relatively unskilled members of the workforce. The requirements of modern society are for a few highly skilled individuals, and for increasingly large numbers of people whose jobs are deskilled or who are unable to obtain jobs at all. As Braverman points out, although the *average* level of skill in society may have increased, a majority of the workforce have suffered a decrease in the skill that is required for their work as individuals — and at the same time, they have suffered a diminution in the control they exercise and the satisfaction they derive from their working lives.

From the point of view of pupils' mathematical knowledge, the effects of this kind of degradation of the work process are not hard to identify. Supermarket checkout personnel operate the most sophisticated computing machinery, but no longer even have to calculate customers' change (let alone understand how the machinery works). Computer operators and most programmers now have largely routinised jobs, less rewarding or intellectually stimulating than a Victorian office clerk. It is hard to draw parallels between jobs across time, but Chris's working conditions are — in some though not all respects — more comparable to those of a modern assembly-line worker than those of a pre-war banking clerk. And though the description of Chris's job is a stereotype, it realistically depicts the relationship which an increasing number of workers have to the productive system, to their fellow-workers, and to others with whom they come into contact.

Braverman's argument belongs essentially to the realms of political economy, and as it stands, it runs the risk of providing an over-deterministic interpretation of the relationship between schooling and society. More particularly, the analysis raises a range of educational questions which are unanswerable from a strictly economic viewpoint — precisely because the economic base does not *determine* the educational superstructure. In fact, much of the specifically educational contribution to this issue in the last fifteen years has been aimed at clarifying the relationship between school and society, and in particular, drawing out the implications of various theoretical frameworks for curricular practice. The seminal work in this field is that of Bowles and Gintis (1976) who established their principle of 'correspondence' between the social relations of production and the social relations of education. Crucially, they argued that the threefold expressed aims of education — firstly to reduce inequalities, secondly to encourage personal development, and thirdly, to integrate individuals smoothly into social and economic life — were incompatible. On the contrary, they made a convincing case that schooling manifestly fails in its first two ambitions — that on the whole it reproduces social inequality rather than the reverse, and that it represses personal development rather than fostering it. Essentially, Bowles and Gintis' conclusion is very much in line with Braverman's: what schools do best is to socialise future generations of young people into the appropriate niches they are destined to fill as adults.

Bowles and Gintis' perspective has been thoroughly analysed and subjected to intense criticism, not least for its reductionist perspective and the pessimistic implications of the theory. This essay is not the place for an examination of these and related issues (for a concise review of Bowles and Gintis' early position, together with an analysis of the work of competing social theorists such as Althusser, Bourdieu, Gramsci and others, see Giroux 1983). While there has been no shortage of critics who have seen in this kind of social theorising a confirmation of their belief in the crude determinism of the Marxist approach, more thoughtful critics have attempted to outline the conclusions and limitations of the correspondence principle, and to investigate the ways in which it is expressed in schools and classrooms. Indeed, as Whitty (1985) points out, criticisms of crudely deterministic accounts have come increasingly from those working within a Marxist tradition.

Current approaches within this theoretical framework are increasingly stressing the educational setting as an arena of struggle and contradiction. Far from simply reproducing the economic and social relationships into which it is inserted, the school has come to be seen as a social site in which teachers and students do not simply respectively transmit and receive information, but in which they also produce and mediate that information. Thus, the critical insight that schools exist in a particular relationship to the economic and social system within which they operate does not necessarily imply that the relationship is merely one of simple correspondence. As Giroux puts it, "while there is little doubt that schools are tied to educational policies, interests and resources that bear the weight of the logic and institutions of capitalism, they also provide room for emancipatory teaching, knowledge, and social practices". (Giroux 1983, p.115).

This contradictory role of the school has important consequences. Primarily, schools are to be seen as not only structured by, but also as structuring agents for, the society which gives birth to them. Viewed from this perspective, schools place particular constraints, but also open possibilities on teachers and their pupils. Perhaps most crucially, this view offers a framework within which to understand how mismatches can occur between the priorities of the system and the realities of schooling, and how the forces which shape curricular innovation are themselves subject to change.

CURRICULUM CHANGE IN MATHEMATICS

The fundamental issue which this brief theoretical survey is designed to illuminate, is to offer some mechanism for describing and understanding the social basis for curriculum change in mathematics, and in particular, how changes in this basis have been reflected in the introduction and subsequent destinies of curricular reforms.

In essence, the general position I want to put forward is this: during the post-war economic boom, the social relations of education at all levels were developed in line with the economy's thirst for skilled white-collar labour to staff accelerating technological innovation. In accomodating to this process, much of the educational practice hitherto reserved for the professional and managerial elite was appropriated for these new layers who, far from being part of that elite, were actually destined to fill the increasingly large but relatively powerless middle levels of the industrial and service sectors. The expansion of UK education for most of the last thirty years has, therefore, been based — far from a simple 'correspondence' between the economic and educational priorities — on a structural mismatch between the system's requirements and the schools' ability to deliver them through the curriculum.

I am all too aware of the over-general nature of the above scenario. What, for example, is 'the system', and to what extent — if any — is the mechanism I describe the result of intentional policy? Rather than answer these and related questions in the abstract, I prefer to focus on a specific example in order to discuss its implications for a more general consideration of curricular change in mathematics, namely the familiar and well-documented introduction and apparent failure of the modern or new mathematics movement initiated in the nineteen-sixties.

The rhetoric of modern mathematics was — in the UK at least — rather straightforward. At root, it hinged on the idea that the display of mathematics as a formal system, unified by the idea of *sets,* would allow pupils to see mathematics as a coherent whole, offer a greater degree of involvement and understanding, and result in a more substantial fraction of pupils studying mathematics successfully at post-school levels. I am not here concerned with the specific role of the set as a unifying theme (in fact, there are a number of other mathematical ideas which could have served equally well), or of the ways in which the pedagogical and structural aims of the reforms came into conflict. I am interested in understanding how the idea of sets *functioned* in the school curriculum, and in the broader society of which schooling is a part.

It is commonplace to recognise the extent to which the introduction of 'modern' mathematics was driven by societal pressures (spūtnik in the US; economic competition with Japan and West Germany in the UK). As a corollary, the reforms resonated with the pressures of the market, and although they fitted with the priorities of mathematicians (and some mathematics educators), pedagogical aims were largely sacrificed to content for its own sake legitimised by mathematicians (Hayden and Rudolph 1984). Of course, this situation is hardly new. The priorities of the mathematics curriculum — certainly in the period up to the second world war — were and still are largely developed by and for a small minority and imposed on the rest.

But the situation is not so clear-cut. A mathematics curriculum based on the structure of the subject fitted the perceived priorities of nineteen-sixties capitalism, a system which was experiencing a massive boom accompanied by technological advances, an improved standard of living, and a general social questionning which overturned many of the cultural and political assumptions of the preceding decade. Chris was being taught set theory. No matter: the image of children gaining an overview of the mathematical system — the better to understand it — resonated nicely with corresponding changes perceived to be taking place in schooling and society as a whole.

It is important to clarify just how little intentionality is built into this scenario. In fact, it is precisely the *lack* of intentionality involved which plays such a critical role in bringing to the surface the contradictions inherent in the situation. For in reality, the economic system of the nineteen sixties was laying the basis for an expansion of jobs like Chris's: far from generating a 'need' for more and more people with an overview of 'the system', for most people, the labour process was becoming more and more deskilled, more automated, and more alienated than ever.

If the majority of pupils are studying mathematical content which seems to be inappropriate for equipping them with the skills necessary for their role in the production process, there are inherent instabilities which will, sooner or later, emerge. At first sight, it is surprising that schools can withstand such pressures at all — after all, if the utilitarian rhetoric underpinning mathematical education is to be believed, mathematics is primarily taught because it is what pupils "actually need to use in later life, particularly at work". (D.E.S., 1988). But in reality, it has been not the *training* but the *socialising* function of mathematical learning which has functioned most consistently. Behind the utilitarian rhetoric in which debate on the content of the mathematics curriculum is increasingly framed, there has been a consistent view of mathematics as performing a function other than simply the transmission of numerate skills to staff niches in the labour process. Howson, Keitel and Kilpatrick (1981) illustrate the long history of this view with a source dated 1857, which advocates mental arithmetic to foster 'the habit of promptitude', arithmetic to induce 'habits of exactness and order', and mathematics to instil the 'acquisition of necessary truths'. (Quoted in Howson, Keitel and Kilpatrick p. 24).

While Western economies were in the post-war boom era of relatively full employment, this contradiction did not rise to the surface. But with the system entering crisis the situation changed radically. Now the social function of education became paramount and explicit at the expense of the educational. James Callaghan, the Labour prime-minister whose 1976 speech on education at Ruskin College foreshadowed so much of what was to become the extreme right's educational orthodoxy of the eighties, stated: "There is no virtue in producing socially well-adjusted members of society who are unemployed because they do not have the skills"' This legitimation of the back-to-basics movement had as its primary aim, the laying of the responsibility for economic decline at the schools' doors. As Aronowicz and Giroux (1987) so aptly put it a decade later, "Industry has rediscovered education because it has lost its once secure markets"' (p.10).

Thus, while not ignoring the pedagogical issues involved, the failure of modern mathematics (and the planned demise of the 'investigative' approach in the UK[1]) can be viewed as an expression of the emerging contradiction between the priorities of education and those of industry. The latter's calls for a return to arithmetic drill may be viewed in this light: "...so called 'back to basics', while having little rationale in terms of either pedagogical or technological reason, may be understood in part as a response to the failure of correspondence between schools and capitalist production brought about by the dynamics of the accumulation process confronting the inertia of the educational structures" (Gintis & Bowles, 1988, p. 20). The very conceptual understanding which the modern movement was attempting (and in a few cases succeeding) to introduce to its pupils, were precisely the skills which the combination of new technology, mushrooming centralisation of industry, increasing international competition, and industrial deskilling (not to mention increasing unemployment) were needed less and less.

Like Chris herself, her predecessors never 'needed' an overview of mathematical structures (or a creative approach to mathematical problem-solving); but Chris does 'need' to fit into the bank's social structure, and to adequately perform her role (and no more) within it.

Of course, from an educational perspective, the process seems rather different — even if it has the same effect. Modern mathematics (like the investigative movement of the nineteen-seventies and eighties, and, more recently, the introduction of computers into mathematics classrooms) turned into something other than the unifying theme it was intended to be by some

1 Ironically, this is now reappearing in a similar form - and for different reasons - in the USA, under the guise of 'constructivism'.

of its protagonists. Sets were manipulated, rules were remembered but little emerged in terms of the hoped-for structural overview which lay behind the attempt. In short, it is possible to view the 'failure' of set theory ion terms of its transformation from a means for children to unify disparate mathematical ideas, to a piece of syllabus content which was an end in itself. However, I do not propose to discuss the shortcomings of the innovations from a pedagogical point of view — there is in any case a sense in which the pedagogical objectives of modern mathematics were never realised (see Hoyles 1988 for a discussion of the issue).

In content terms, much of the mathematics curriculum can thus be viewed as an example of the way in which school mathematics stands dissociated from its connections with science and technology and divorced from its applications not only to the sciences but to reality — a process which was begun and effectively completed in the 19th. century. "This sort of mathematics could stand beside Greek and Latin as the instrument of a coherent body of educational ideology and practice" (Damerow and Westbury, 1985). Thus mathematics becomes part of what Bourdieu has termed the 'symbolic violence' exercised by schooling, in which groups seek to impose and legitimate meanings while disguising the power relations which are derived from them (see Brousseau 1984 for a provocative analysis of the teacher-pupil relationship within a mathematical context). This process, when taken to its conclusion results in the process which Postman provocatively identifies in the domain of literacy, by suggesting that control is exercised by "carefully discriminating against what the student knows — that is by labelling what the student knows as unimportant" (Postman 1970, p.251).

THE SOCIAL DYNAMICS OF THE MATHEMATICS CURRICULUM

The curriculum in liberal democracies is in disarray. The clash between the needs of the individual and the needs of the system is resulting in mass disaffection and truancy. Schools are blamed for all kinds of social ills, and teachers are attacked for imparting 'dangerous' ideas to their pupils. It is tempting to laugh off such attacks as simply an attempt by political commentators to shift the blame from themselves onto teachers. But there is a grain of truth which is recognized by the right, and often ignored by the left:

> ...since education can also stimulate questioning and criticism, it is hardly suprising that some governments have been tempted, or forced, to seek direct control over educational content, preventing the imparting of 'dangerous knowledge' (Crompton and Gubbay quoted in Alexander, 1991, p. 73).

There are a number of questions which arise. To what extent can we include *mathematical* knowledge as 'dangerous knowledge'? Can we really argue that knowledge of, say, how to solve a linear equation threatens the status quo? Does it make sense to argue that *all* mathematical knowledge is potentially 'dangerous' or is some more dangerous than others? And in any case, we need to ask to whom it may be dangerous. Is there mathematics which challenges the status quo to be counterposed to that which does not?

It is clear that some *do* believe that there are distinctions to be made among mathematical knowledge. The recent battle with various UK government agencies which resulted in the downplaying of 'using and applying' mathematics and the introduction of long division algorithms illustrates nicely that those in power — and those resisting such power — do *not* believe that mathematical knowledge is politically neutral. At the rhetorical level, the debate was about downplaying an area of mathematics seen (by both sides) as potentially enlivening mathematical activity, opening access to mathematical thinking and offering opportunities for mathematical creativity. On the other hand, the introduction of long division represented an attempt to impose on the curriculum a piece of mathematical knowledge which is both anachronistic and *use*-less. At precisely that point in the evolution of human knowledge when knowledge of how to divide a three digit number by a two digit number becomes redundant, the government of the UK passes a legal requirement to teach it to all pupils[2].

Of course, one needs to distinguish between rhetoric and reality. Indeed, it is not altogether useful to conceive of curricular policy as if it were planned by tightly-knit cabals with clear and unambiguous objectives. Nevertheless, trying to make sense of situations such as this leads us to at least question the motives of those involved. In whose interests *is* it that children learn long division, rather than focus on how mathematics is 'used and applied'? In what sense, if any, is the popular imagination right in thinking that 'basic skills' are a prerequisite for individual prosperity and national recovery? What are we to make of the argument that the well-being of the nation depends on the inclusion of the long division algorithm, but the exclusion of its use and application?

These kinds of questions clearly illustrate the limitations of Bowles and Gintis's earlier position. If there were some simple 'correspondence' between the needs of the economic base and the shape of the educational superstructure, we might expect the mathematics curriculum to be reduced to the basic requirements of numeracy, but we would not necessarily predict the reintroduction of archaic algorithmic procedures. In any case, the process of denuding the curriculum of mathematical content which is proceeding apace in the UK[3], is taking quite the opposite direction in the US, where there seem to be some indications that bottom-up proposals to broaden and mathematize

---

2 Actually, only those who are not privately educated.

3 Paradoxically, by testing (from age 7) hierarchies of tightly-specified 'statements of attainment'.

the curriculum are taking place: the UK and the US economies may be at variance in terms of prosperity, but their fundamental social and economic relations of production are the same.

However, Bowles and Gintis themselves have indicated a way out of the dilemma. From their point of view, the issue is not so much the *content* of what is taught, but the form:

...it is the FORM of the educational encounter — the social relations of education — that accounts for both its capacity to reproduce capitalist relations of production and its inability to promote ... healthy personal development. The actual CONTENT of the curriculum has little role to play in this process. (Gintis and Bowles, 1988 p. 28).

I think that Gintis and Bowles go a little too far in stressing the relative importance of form over content. The lines around mathematical content may be arbitrarily drawn, but they are not irrelevant. Indeed, it is because the social relations of education are (in part) mediated *through curriculum content* that innovations (such as buildling a curriculum on set theory) often end up functionning in unintended ways — in Chevellard's words they become 'didactically transposed'. So for mathematics, the importance of long division is not to learn long division, but in what long division stands for: tedious, repetitive, routine tasks. In the minds of some, 'using and applying mathematics' stands for creativity, intellectual challenge and innovatory thinking[4]. The question is not whether the particular content determines its social function (it does not), but how content and social function relate to each other.

The exact nature of curricular content may be secondary in achieving the 'aims' of schooling from the point of view of reproducing 'capitalist relations of production', but it is not secondary for others engaged in the education process. On the contrary, I think that curricular content — often relegated to a secondary position by social theorists of the curriculum — needs to be reasserted in the light of attempts to suppress it:

...most sociologists of education have completely neglected the role of thinking-tools for critical consciousness. They all join the large group of pedagogues who claim that pedagogy is about human beings, not subjects....some of the deepest thinkers in the field of sociology of education to some extent disregard the functionality of knowledge as a means for a democratic education. (Mellin-Olsen 1987, p. 205-6).

Mellin-Olsen is arguing that knowledge is functional, and by implication, that those who argue — either in celebration of 'process' or to prioritise routine skills over understanding — for its downgrading are covertly or overtly arguing for an uncritical educational system. But it would be a mistake to consider that the priorities of the powerful always reside in

---

4 I do not question here whether they are justified in this view.

restricting curricular content, or that the powerful constitute a homogenous group with a uniformity of aims for education. For example, in the US currently, many industrialists are arguing — as they did in the UK during the 'white heat of the technological revolution' — that there is a need for more than basic skills, or rather that basic skills involve social as well as cognitive facets:

Beyond good basic reading, writing, and computation skills, employers expect competence in creative thinking, personal management, and interpersonal relations. Also critical are abilities to organize and verbalize thoughts, conceptualize, resolve conflicts, and work in teams. (in Lewis & Gagel, (1992) p. 134).

It is not altogether clear to what extent sentiments such as these run counter to the pressure for deskilling the learning process and the removal of content from the curriculum. There are, to be sure, powerfully conflicting tendencies among employers of labour who do not speak with one voice on educational questions, any more than they do on economic and political matters. There are also those who are realising the productive limitations of Fordist labour policies; surprisingly but most significantly many Japanese companies in Britain have adopted working practices completely at variance with the alienating, monotonous and demeaning (as well as increasingly uncompetitive) work practices so well described by Huw Beynon's (1973) book *Working for Ford.* Interestingly, Mathews (1989), in his book *'Tools for Change'*, argues that certain sections of industry are moving towards a post-Fordist view which challenges the assumption that "it is most productive to vest intelligence and control in technical systems, and treat workers like donkeys". He suggests that the notion of computers acting on behalf of workers is being superceded by the construction of computers which work alongside them, allowing workers to exercise traditional craft skills and to exercise certain (restricted) forms of control over their working practices.

It seems entirely possible that there will be found new, more productive, ways for Chris to work alongside the bank's computer system. It is too early to say whether these will constitute a 'post-fordist' approach which will seriously undermine the social dynamics of the system or necessitate a further shift in educational policy. It may be that this industrial voice is louder across the Atlantic than it is here, and that this in part explains the changes taking place in the US. Nevertheless, whatever the eventual shape of industrial working practices, it would be surprising if the mathematics curriculum remained immune to their influences.

CONCLUDING REMARKS

Each society (or to be more precise, those in control of each society) believes that the mathematics curriculum is part of the recognised corpus of a society's knowledge and culture and that as such, it plays a crucial role in determining what students know, what they bring to that society when they leave school, and — most critically — how committed they feel towards it. Of course, there are many ways in which social theorists have conceptualised the social determinants of curricula. Nevertheless, in the fifteen or so years since Bowles and Gintis's book, there has arisen general agreement that the curriculum is not *determined* in any general sense; that is, there is no *necessary* shape to the curriculum (mathematical or otherwise) which arises from particular interests, societal needs or the balance of social forces. On the other hand, it seems worthwhile to try to make sense of the relationship between the mathematics curriculum and the social, cultural and demographic trends in society. I would like to try and understand how Chris's beliefs about, attitudes towards and understandings of school mathematics relate to those she holds in relation to her working life.

At the risk of labouring this point, I will draw an analogy. Why do human beings have two legs? To ask the question is to misunderstand the nature of evolutionary theory; human beings do not have two legs for any *reason,* because it was planned that way, because two legs are best (note, however, that a common abuse of language *does* often conceptualise evolution in just such a way). We can, however, answer a different question: namely, how is it that two legs works out satisfactorily — so satisfactorily in fact, that human beings evolved this way? How does two-leggedness *function?* Answering these questions involves describing the ways in which two legs suit human beings (not that they *couldn't* have three legs), and the ways in which the stability (in evolutionary *and* mechanical terms) of two-leggedness makes it unlikely that selection will evolve towards some other number of legs.

And so it is with educational change. Chance and inertia both play their part; what actually *is* is partly a matter of chance encounter, the conjuncture of (for example) political events or even politician's whim; and change is difficult, selected out — a critical determinant of how the curriculum looks now is how it looked yesterday. It may not be altogether useful to chart each chance encounter with political reality, or to describe in detail the inertial frame within which educational change occurs; it may, instead, be more productive to map some of the ways in which the mathematics curiculum functions within society, and to discuss the possibilities for change within it. This is no easy task, not least because of the impossibility of standing outside the system in order to view it more critically: our ideas about mathematics and education are, in general, not 'created', they form part of ideology which makes our social and intellectual world intelligible to us. However, ideology also serves to blind us to the ways in which meanings are created, the mechanisms by which 'truth' is justified and legitimated. And there are, of course, lies as well.

REFERENCES

Alexander, D.: 1991, Curriculum turmoil in liberal democracies. *Journal of Curriculum Studies* p.71-78.

Aronowicz, S. and Giroux, H.: 1987, Ideologies about schooling: rethinking the nature of educational reform. *Journal of Curriculum Theorizing* 7, 1, 7-38.

Beynon, H.: 1973, *Working for Ford* Harmondsworth: Penguin.

Bowles, S. and Gintis, H: 1976, *Schooling in Capitalist America.* London: RKP

Braverman, H.: 1974, *Labor and Monopoly Capital: The Degredation of Work in the Twentieth Century.* New York: Monthly Review Press.

Brousseau, G.: 1984, The Crucial Role of the Didactical Contract in the Analysis and Construction of Situations in Teaching and Learning Mathematics in Steiner H. (Ed.) *Theory of Mathematics Instruction.* IDM: Bielefeld.

D.E.S.: 1988 *Mathematics for ages 5-16.* London: HMSO.

Damerow, P. and Westbury, I.: 1985, Mathematics for all - problems and implications *Journal of Curriculum studies,* 17, 2, 175-184.

Gintis, H. and Bowles, S.: 1988, Contradiction and reproduction in educational theory in Cole M. (Ed) *Bowles and Gintis Revisited,* p. 16-32. London: Falmer.

Giroux, H.: 1983, *Theory and Resistance in Education* London: Heinemann.

Hayden, R. and Rudolph, W.: 1984, Will there be a New 'New Math?' *Journal of Curriculum Studies,* 16, 3, 311-316.

Howson, G., Keitel, C. and Kilpatrick, J.: 1981, *Curriculum development in mathematics* Cambridge:Cambridge University Press.

Hoyles, C.: 1988, Paradoxes in Mathematics Education in the 1980s In: Clough et al. (Eds.) *The new learning: change and challenge in curriculum design.* Macmillan.

Lewis, T. and Gagel, C.: 1992, Technological literacy: a critical analysis *Journal of Curriculum Studies* 24, 2, 117-138.

Maclure, J. S.: 1965, *Educational Documents.* London: Methuen.

Mathews, J.: 1989, *Tools of Change: New Technology and the Democratisation of Work.* Marrickville NSW, Pluto.

Mellin-Olsen, S.: 1987, *The Politics of Mathematics Education.* Dordrecht: Reidel.

Noss, R.: 1989, Just testing: a critical view of recent change in the UK Mathematics Curriculum. in Clements K. and Ellerton N. (Eds.) *School Mathematics: the challenge to change.* Deakin University Press.

Noss, R.: 1990, The National Curriculum and Mathematics: a case of divide and rule? in Dowling P. and Noss R. *Mathematics versus the National Curriculum.* London:Falmer Press. pp. 13-32.

Popkewitz, T. and Pitman, A: 1986, The idea of progress and the legitimation of state agendas: American proposals for school reform. *Curriculum and Teaching,* 1,1, 11-24. TES, 3.6.88, p17.

Postman, N.: 1970, The politics of reading. *Harvard Educational Review,* 40.

Whitty, G.: 1985, *Sociology and school knowledge.* London: Methuen.

Dr Richard Noss,
Department of Mathematics, Statistics & Computing,
Institute of Education,
20 Bedford Way,
London WC1H 0AL U.K.

J. VOLMINK

MATHEMATICS BY ALL

INTRODUCTION

The argument in this chapter is that mathematics, even before its professionalisation, has always been the domain of the select few. Attempts have been made, in recent times, to challenge the Eurocentric bias in mathematics and this has led to a greater appreciation of the mathematical contributions of different cultures. While there is an ackowledgement that mathematics is a pan-human activity, there is no evidence either in the history of mathematics or in mathematical practice today, to support the belief that, within a particular cultural context, mathematics was widely practised by the majority. The social arrangements of early civilisations were such that only the rich, the powerful, the influential, had access to mathematical knowledge. At times there was almost a conspiracy to keep the codified mathematical knowledge as secret as possible. Since there was no mass schooling until about a century ago, this kind of knowledge was only passed down within a certain 'brotherhood'. The fact that state schools have now become a given in most societies offers us the unique opportunity to make mathematics accessible to all. Yet, in spite of a century of mathematics instruction, most people still feel alienated from the subject. In this chapter I argue that we need specific strategies to address this in order to encourage those who have been traditionally under-represented to participate in the production and use of mathematical knowledge. In particular, there should be a shift from seeing mathematics as involving the "interpretation of symbolic information" to an emphasis on situating it in the realm of everyday experiences of people.

THE IMAGE OF MATHEMATICS

Mathematics has been mistakenly regarded as the sole creation of a few, singularly brilliant (white male) individuals. This has placed mathematics outside the realm of experience of most people . It makes a sharp distinction between the few and the many as a fundamental assumption of social organisation. In other words, having disinvited participation by the majority, the practice of mathematics now gives a lot of power to a minority. It is clear therefore that the maldistribution of mathematical experience is not without political implications and moral consequences

Mathematics is not only an impenetrable mystery to many, but has also, more than any other subject, been cast in the role as an 'objective' judge, in order to decide who in society 'can' and who 'cannot'. It therefore serves as

S. Lerman (ed.), Cultural Perspectives on the Mathematics Classroom, 51–67.

the gatekeeper to participation in the decision making processes of society. To deny some access to participation in mathematics is then also to determine, *a priori* , who will move ahead and who will stay behind. Furthermore, there is a certain oppressiveness about the role of numbers and mathematics in society. This is felt at both personal and collective levels. A climate and a culture of resistance towards mathematics have developed in society. People are generally not neutral in their affective commitment to mathematics, but rather feel extremely alienated from, if not actively hostile to it. In fact, it has become almost a status symbol to display one's disdain or disinterest in mathematics.

CHANGING THE PRACTICE OF MATHEMATICS

I also believe, perhaps naively, that mathematics can help all of us enter more fully in the choices in our lives and that the world would be a much better place if people were able to participate in the choices that affect their lives. To the extent that mathematics explains things, it can help us to examine ideas that we otherwise could not, and to create fresh ones. It does help us to put a structure to our experience of our world, to articulate our images and ideas about that world and to see the contradictions in that world. To know and to understand is a basic human right and, if this right is being denied, at least in part, by the way mathematics is seen and taught, then there clearly exists a need to democratise mathematics. This means, in the first place, that mathematics needs to be demystified. Demystifying mathematics does not mean merely making it accessible to all, but has also to mean that those who have been marginalised, disinvited and under-represented, will come to see that they too can share in the creation of mathematics and hence in its ownership, beauty and power. In other words, it is simply not good enough to equate 'democratising' mathematics with revealing the once hidden mathematical 'discoveries, skills and facts' to people on a broader scale. Of course this is necessary and important. There is a body of conventionally understood and publicly accepted mathematical ideas and skills which has been culturally negotiated over centuries and it is vital that the maldistribution thereof be addressed. But skills and facts are not, and should not be seen as ends in themselves; neither do they have an unproblematic status. There are strong hegemonic forces at work in society, that impose a certain view of mathematics on us all. Our schooling has in many ways encouraged us to accept as unproblematic, that the traditional mathematics curriculum somehow embodies uniquely powerful knowledge and eternal truths which should be taught and learned in a catechistic fashion. Furthermore, this draconian body of knowledge is not only infallible but also*universal.* Mathematicians have for too long regarded their discipline as a set of bounded meanings, with a well-defined, culturally-neutral and value-free, inside and a large outside. They, as high priests, have the responsibility and sole right to decide what should be included and what should be excluded. It is easy to see why mathematics is taught and learned at school in a context sadly lacking in skepticism. Mathematical 'facts' and 'skills' are being pursued with such vigour, and sometimes even with sincerity, that one would almost like to believe that these, in themselves, are sufficient to constitute mathematical competence.

It is often said: "If you can make it, you can own it." If this is true for mathematics then the notion of ownership is intimately connected to the idea of construction. *Mathematics for all should therefore become, mathematics by all.*

MATHEMATICS AS HUMAN ACTIVITY

No matter how far back we go in history, we will find that every culture, in all places, was engaged in some kind of mathematical activity (Bishop 1988). Much of this activity included counting, measuring, locating, classifying and orienting. These activities and many others often grew out of practical necessity, but also had strong connections with aesthetic considerations and ritualistic practices.

In all of these activities, geometry always seemed to have played a vital, and at times, a central role. This may be because of the intuitive human ability to recognise and compare shapes and sizes of physical forms. For example, some forms are regular while others are not, some have an inherent symmetry, while others do not, some are large and some are small. The cumulative daily experiences and observations of human beings since earliest times, must have invited the inquiry of a reflective mind. The concepts brought to light from these observations were used in art and daily activities since earliest times. Eves (1963) refers to these activities as "subconscious geometry" (p. 2).

Most textbooks on the history of mathematics such as Boyer (1968), Bell (1945) and Kline (1972), present the traditional view that geometry first became an organised body of knowledge as a direct result of the early Egyptian agricultural activities in the Nile basin. Eves (1963) claims that similar agricultural as well as engineering activities also occurred in the other great river basins such as the Tigris and Euphrates of Mesopotamia, the Indus and Ganges of south-central Asia, and the Hwang Ho and the Yangtze of eastern Asia. The geometry which was used by these societies to erect great structures, irrigate, drain and bound their land, is sometimes called scientific geometry (Eves, 1963).

Seidenberg (1960) in a paper entitled 'The Ritual Origin of Geometry," presents an alternative view on the 'origins' of geometry. He puts "theologic-geometry" (p. 520) before Egyptian and Babylonian geometry. He argues that geometric rituals, such as the building of temples and altars, can be traced back as far as 4000 B.C. In this case the priests, who were the architects of the temples and altars, were the only geometers.

Both of these views raise a fundamental objection in my mind. They both attribute the creation of mathematics and in particular, the origins of geometry to the rich and the powerful. The educated priests and the skilled land-owner were, in these views, almost uniquely endowed with the insight to conceptualise mathematical ideas beyond the level of everyday activities. But geometry did not suddenly appear, it was not discovered by a few elite. It was created by people and was accessible to the masses from the earliest of times. It is what I would call 'people's mathematics.' People were building dwellings long before they were building temples or bridges. Some of these were round while others were rectangular. These were not just random choices; they were reasoned and well-founded on a body of geometric knowledge (Zaslavsky 1973). Although I am willing to concede that the Egyptians made a considerable contribution to geometry, I am much less willing to admit to a thesis which locates the origin of geometry at any one place at any one point in history.

For the Babylonians, geometry did not exist as a separate discipline but was integrated into all their arithmetical procedures. It is true however, that geometry as it appears today, is very different in character from the geometry known to the Babylonians or the Egyptians. As geometry passed from the Egyptians to the Greeks so also did its character change.

Eves (1969) puts it thus:

> The Greeks insisted that geometric fact must be established, not by empirical procedures, but by deductive reasoning; geometrical truth was to be attained in the study room rather than in the laboratory. In short, the Greeks transformed the empirical, or scientific, geometry of the ancient Egyptians and Babylonians into what we might call "systematic," or "demonstrative," geometry. (p. 171)

The Greeks appear to be the first to have introduced the notion of 'formal' proof into geometry. But, for Eves to assert that this was the first example of "systematic geometry,' is not quite accurate. In fact, there is abundant evidence that ancient Egyptian and Babylonian geometry was more than a mere collection of unsystematic empirical procedures. For example, the ancient Hindi, *Sulbasutram* of Babylon, was a systematised body of geometric procedures used to construct temples and altars. The fact that this could be seen as demonstrative geometry, is particularly significant in the light of the claim by Seidenberg (1960), that this document dates back to about 4000 B.C. The nature of (what became known as) geometry did take on a more 'formal' appearance in Greek times. Appeal was made to geometry to justify algebra and analysis in a 'formal' way. Geometry was the standard of rigor and the other branches of mathematics did not have their own innate formal systems until the end of the nineteenth century. The Greeks undoubtedly raised geometry to a level of great prominence in their mathematics. Wilder (1968) suggests that one of the reasons for this was because geometry associates the most basic element of mathematics, namely, *number,* with the concept of *line,* one of the most basic elements of geometry.

The remarkable thing is that the new element, geometry, so completely took over mathematics, at least methodologically, in the Greek culture. Presumably the elevation of geometry to a dominant position resulted chiefly from the manner in which geometry could be used to discover and prove number-theoretic theorems, as well as to cope successfully with the general real number in the guise of "magnitude."(p. 95)

To assume however, as some do, that Greek mathematics is 'pure mathematics' while Egyptian geometry is 'applied mathematics,' is to greatly misinterpret the nature of pre-Hellenistic mathematics. Mathematics until the time of Euclid was always intimately connected to human activity. No contradiction was seen between human purposes and human knowledge, between pure and applied, practical and theoretic. Long (1986) aptly points out that the distinction between pure and applied mathematics as a tool is *read into* the history of mathematics. It conceals the problematic character of the distinction and of our narrow conception of 'toolhood'.

.... we often read that mathematics began in ancient Egypt as applied mathematics. Indeed, this is shown by the very name: "geometry" means earth measurement. (Of course this name is Greek, not Egyptian, and reports the Greek understanding of what the Egyptians were doing.) To call an activity "applied mathematics" is to appeal to our distinction rather than to show it is applicable. That distinction is based on the possibility of using mathematics as a tool with which to understand and manipulate nature. But this possibility exists only within the modern Western conception of nature, a conception which appears to be totally foreign to the ancient Egyptians, the ancient Greeks, and even the medieval Christians.(p. 612)

In fact if we accept the arguments by Lakatos (1978) that mathematics is quasi-empirical in nature, we will have very little difficulty in accepting the geometry before Euclid as valid mathematics. Euclidean geometry, however, was not able to stand the test of time as the only model capable of representing physical space. It is interesting that this very model to which appeal was made to formalise mathematics, was also the very reason why the foundations of mathematics were shaken in the middle of the nineteenth century.

With the advent of a different, yet consistent, hyperbolic geometry, the limitations and non-uniqueness of the Euclidean model became apparent. The present-day view of geometry is not of a unitary one but one of several geometries. There is non-Euclidean geometry, differential geometry, projective geometry, topology and so on. Geometry therefore has more than one meaning. Many mathematicians would go further to deny that geometry exists as an independent body of knowledge. Meserve (1973) for example, states:

...geometry as such is no longer a subject. What is important is a geometrical way of looking at a mathematical situation -*geometry, is essentially a away of life.* We have geometric topology, geometrical dynamics, differential and algebraic geometry, but not just 'geometry.' (p. 243)

While I have some difficulty with the non-existence of geometry, I like the notion of regarding geometry as a way of seeing. Seeing things geometrically allows one to see things that you might not otherwise have seen and enables one to create thoughts that you could not otherwise have conceived. While geometry is not a very good example of a formal mathematical system, it does provide us with ways of looking meaningfully at all areas of mathematics. Geometric interpretations will continue to hold a connection with our intuitions in a way that no other branch of mathematics does. It calls for a very different set of actions and representations than does any other mathematical activity.

One of the greatest powers we possess as human beings is the power to imagine. We can create images of things both physical and abstract. Caleb Gattegno (1965) describes geometry as an 'awareness of imagery.' Images involve much more than thinking about things statically. It arouses a whole complex set of feelings and of mental actions. Such imagery is a dynamic process of the mind.

Whether or not geometry started off as measurement of the earth, it nevertheless has now outgrown its name. While still retaining its connections with human activity, that activity goes beyond simply measuring lengths, areas and volumes. It has become a rich mental and psychological activity which involves the creation of ideas and the validation thereof. It involves the construction of images of abstract notions. It is dynamic and it is exciting. It is indeed a worthy intellectual pursuit.

The purpose of going into this detail about the development of geometry, is to illustrate how a particular branch of mathematics cannot be separated from its roots in everyday human activity. Similarly with other branches of mathematics, some of which have since been formalized and professionalised beyond recognition, we can always trace their rudimentary origins to some fundamental aspect of human activity. In his book Mathematics for the Millions, Lancelot Hogben(1968) describes the evolution of the language of measurement and counting. He describes how this language emerged out of the changing social circumstances and achievements.

As human inventiveness has turned from the counting of flocks and seasons to the building of temples to the steering of ships into chartless seas, from the seafaring plunder to machines driven by the forces of dead matter, new languages of size have sprung up in succession. (p. 28)

So mathematics did not just arise out of the world of our sense experience, but also out of our inventions, our economic arrangements, our religious beliefs and cultural arrangements. My argument is that there is almost no evidence in school mathematics that mathematics has any bearing on these issues; in fact it has been almost completely dislocated from children's reality in particular. I am not suggesting that we should find possible applications for every single mathematical concept encountered in school, but rather that it would be much more intellectually honest if we should re-locate mathematics within the discourse of everyday experience and thus show that mathematics, contrary to its image, is not removed from life, but rather helps to explain it. It is one way of putting a structure on our experience.

MATHEMATICS IN SCHOOLS

I have turned to the history of mathematics in the hope of finding some evidence that mathematics can be an activity for the masses, by the masses. But this quest has left me comfortless. No matter where I look, I can only find confirmation that only the elite, the select few were and still are involved in mathematics. Now, with the advent of ethnomathematics came also a recognition that the monolithic voice of Euro-centric mathematics cannot be left unchallenged. Ethnomathematics is therefore one of the voices that have come up in revolt against the suppression, lack of recognition and exclusion of the mathematical ideas of other cultures. Yet strangely, when I appeal to the literature on ethno-mathematics, I see again only the rich, the influential, the powerful, the privileged, having direct access to and control of mathematical ideas, in their own cultural contexts. This meant, for example, that the priests and only those that they favoured, acquired the knowledge to design the temples and altars in all their splendour and mathematical complexity; it also meant that the rich landowners developed the language of measurement and size; the powerful, unique few amongst the Incas knew how to keep the code of the quipu and so on. Of course, today the players have changed but the maldistribution of access and intellectual goods, have not. Who are the people, today, that make economic decisions, that control the mode of production, that have the technological expertise? The fact that they are largely white and male, is no accident of history. It is a clear statement that our socio-political arrangements have systematically de-skilled some and privileged others. A glance at the tables in the Appendix to this chapter illustrates the case with reference to the situation in South Africa.

Power refers to relations among individuals or groups based on social, political, intellectual and material assymmetries that have been created by the structures of society. These asymmetries, across gender, race and class, are constantly reproduced, not in the least by one of our major institutions of reproduction, namely the school. We, as educators must challenge this and change the way our educational institutions operate. I say this because I have nothing else to pin my hope to for a better society, a society in which everyone will become full participants in the choices that affect their lives. What I am saying is that the emergence of state schools within the course of the last century or so, despite all the flaws, holds the strongest hope for democratisation of knowledge. In particular, it is the one means by which we can hope to have a mathematics for all and by all. This has far reaching implications for the pratice of teaching and learning in schools. Among them are:

1. *Provision of a supportive environment in which students will feel free to risk their tentative, intuitive ideas and courageous attempts at performing actions on mathematical problems.*

This will mean, in the first place, that we will need to provide students with the opportunity to give expression to their personal constructions. Such opportunities can only be created in classrooms where there is an atmosphere of tolerance for the diversity of approaches to a problem. This will lead to a co-operative interaction pattern where students and teacher together will work towards the co-ordination of their multiple representations. What is sadly lacking in the practice of mathematics, is a tradition of respect for both formal and informal ways of knowing and proving. Mathematics teachers should be facilitators first, referees second and judges last. The reality is that most students do mathematics in an environment where they come to accept, after some agonising and sometimes devastating experiences, that the only thing that counts is what the teacher wants and what the teacher knows. Their own curiosity gets hidden and they resort to an approach where rules are internalised and followed for no good reason at all. A total lack of skepticism prevails in mathematics classrooms.

I believe that students can and do enjoy the creation of mathematical ideas and that they prefer to acquire concepts by active participation instead of merely by receiving ideas. It is unfortunate that most teaching, particularly at high school, does not allow for this kind of active learning. Direct instruction remains the pervasive mode of instruction in mathematics and it is little wonder that most students find mathematics a dull and unstimulating subject. By contrast mathematicians often express great excitement about their work. Mathematics, they claim, is eventful, compelling and creative.

> In research we see a lot of geometry, a lot of data, a lot of science, a lot of computation - together with more traditional mathematical tools. We see investigation, exploration, and a continual search for pattern. Contemporary mathematics compels attention. It has the power to excite the best minds of our youth and to stimulate renewed creativity in teaching mathematics.

> But this is not the mathematics taught in typical school or college classrooms. Far too often, mathematics is a freeze-dried mathematics -- rigid, cold, and unappealing. Instead of exploration there is drill; instead of investigation, imitation. From elementary school arithmetic to college calculus, mathematics in the classroom is dramatically different from mathematics in practice. (Steen, 1988, p. 418)

The above quote from an eminent mathematician serves to illustrate the disparity that exists between mathematics practice and mathematics in the classroom. There is not the same event sense in mathematics in schools as there is in the practice of the discipline. Part of the reason for this is that learning and teaching are not organised around students' experiences. There may be some experiences provided for students, but whose experiences are they? If we hope to encourage mathematics for all and by all, it will be vital to arrange teaching and learning around authentic experiences of students. I am not suggesting that some mathematics cannot be learned through direct instruction. On the other hand what I am trying to discourage is the location of mathematical learning and teaching at an arbitrary level of abstraction which has no relevance or meaning to students. The cumulative experiences of students have led them to adopt the view that mathematics is the necessary outcomes of meaningless things. This view is not only supported by the culture of learning in schools, but also by the dominant modes of teaching. Meaning does not exist "out there in the world" but is artifactual, that is, it is created and controlled by human beings. Meanings do not reside in mathematical theories, textbooks and journal articles. These are only potentially meaningful. Meaning is created by the direct experience of the individual and in negotiation between individuals. When we ask students to "take charge of their own experience", we are suggesting that they write their own script of meaning. When we place students in a situation which forces them to take *action*, we help them to externalise their thinking. Their efforts are controlled by their pattern of meaning and the meaning is controlled by the context. The focus on meaning draws attention to the powers students possess to make their own sense out of their mathematical experience. It also highlights the conflict between students' own meanings and those formal meanings which are imposed on them from above. Teaching and learning, in this context becomes efforts after meaning. Teaching is the act of sharing meaning, while learning has to do with constantly re-organising existing patterns of meaning. Here we accept the caution given by Confrey (1988) to consider the power relationships which are in operation, when we talk about the 'sharing of meaning'. In other words, we should not take for granted that there is never a threat that students might accept the teacher's meaning instead of genuinely sharing their meaning with the teacher and with each other.

We also need to be careful about the way we refer to 'student experiences' as if these are ends in themselves. Experiences are vital for a democratic mathematics, but on their own they are rather pointless. What is equally important in the mathematical process, is the ability to create mathematical constructs by reflecting on these experiences and for students to defend their constructions and their beliefs. This is what mathematising is all about.

Although others have earlier used the term (Wheeler,1983), I define **mathematising** as a process which finds its origins in an active interaction with our world when we act purposefully and with awareness towards the achievement of certain goals. These goals include understanding of the physical world and acting on it. Mathematics helps us to place a structure on our experience of that world. But it also means understanding the socio-political realities which impact on our lives. Mathematics can make us aware of the contradictions in society and the underlying assumptions of social organisation. Paulo Freire draws our attention to the need not merely to adapt to the world, but also to transform that world. A lot of mathematising also leads to creating new ideas, new perspectives, insights, images and models.

Mathematics is constructed by performing actions on problem situations and then through a process of reflection, a structure is created within which the problematic can be understood and explained (Confrey 1987). In the physical world, a *problematic* may involve analysing behaviour such as: motion, chaos, resonance, stability, convergence, oscillation. *Actions* could involve: represent, control, prove, discover, apply, model, experiment, classify, visualise or compute (Steen,1990). On the other hand, Bishop (1988) uses activities such as: counting, locating, measuring, designing, playing, explaining, to describe actions. As we engage with the problematic through actions, we are forced to *reflect*. Attitudes such as: wonder, meaning, beauty, reality, may impact on our experience. On reflection, attributes such as linear, periodic, symmetric, continuous, etc., may suggest themselves or even dichotomies such as: discrete vs continuous; finite vs infinite; exact vs approximate. In the process, mathematical structures such as algorithms, functions, ratios etc., may be constructed to help us in understanding the problematic.

So we begin with the recognition that situations co-construct knowledge. This means that what we know is inextricably related to how we came to know; in other words, **knowledge is situated** (Seely-Brown et al, 1989). The following table from Seely- Brown et al (1989), shows how students' work differ from that of practitioners and JPFs (Just Plain Folk). For example, whereas practitioners as well as JPFs constantly have to deal with ill-defined and emergent problems and dilemmas, students are asked to act on well-defined problems where meaning is fixed. This leads one to wonder what purpose school is serving.

	JPFS	Students	Practitioners
reasoning with:	causal stories	laws	causal models
acting on:	situations	symbols	conceptual situations
resolving:	emergent problems and dilemmas	well-defined problems	ill-defined problems
producing	negotiable meaning and socially constructed understanding	fixed meaning and immutable concepts	negotiable meaning and socially constructed understanding

Most school mathematics textbooks are written in a style which emphasises drill and practice or routine exercises. At the end of these exercises, some space is given for problems for which a standard recipe for obtaining the answer cannot be used. These 'problems' are generally de-contextualised. At best, they are applications of a previously learnt principle or concept. So application problems in school textbooks are used mainly to provide exercise in, or to illustrate one or other mathematical technique. Some of these applications are more believable than others, but they generally lack authenticity and their educational value must be called into question.

Unlike application problems, situated problems are those that can be seen as emerging out of situated human activity and do not merely serve as illustrations to illuminate a particular mathematical strategy. Mathematical problems should be situated in order to be meaningful and authentic. Situated problems arise outside of formal mathematics, but give rise to mathematical actions, structures and insights.

2. We should not only give attention to how mathematics is learned, but also seriously re-evaluate ***what*** *is learned.*

I am convinced that society cannot continue to afford the luxury of sending children to school to pursue knowledge simply for the sake of knowledge. In other words, the socio-economic and political realities are such that, students must necessarily pursue knowledge for life. So, we must require that school mathematics becomes "**mathematics for life**", instead of a fund of de-contextualised esoteric, abstract and useless knowledge. I am rapidly developing the view that one cannot hope to be transformative by remaining confined to arbitrary subject-matter boundaries. I can therefore envision a future scenario where we finally have to give up our narrow ownership of the various disciplines which we so jealously guard and treasure in order to adopt a collaborative, coherent and integrated approach to education. I have already comes to terms with the desirability of such an approach to curriculum, but realise, at the same time that there are several conceptual and practical difficulties in this regard that first need to be addressed.

However, an honest look at the concrete realities, leaves one with little doubt that, although the arbitrary subject divides and compartmentalisation have not served us well, there are good practical reasons for their existence and that these boundaries are unlikely to disappear in the near future. We therefore have to accept that school mathematics will continue to exist as a separately taught subject for a very long time and that we therefore need to work actively towards transforming curriculum and instruction so that these come in line with the goals of society.

In order to justify its existence as a school subject, the curriculum of any school subject should satisfy at least four goals. viz., personal development, social development, and utility, as well as providing a basis for further learning. Mathematics, maybe more than any other subject, has focused largely on further learning or the professional goal. There is also a long held belief that mathematics must be given a large share of school time because it provides skills that are important in everyday life and work. I am extremely cynical of this view. My own cynicism is rooted in the observation that school mathematics curricula, in its canonical form, although extremely unlikely to be equally appropriate to people in all contexts are so remarkably similar all over the world. I believe that there are strong hegemonic forces at work in the global society that produces this resistance to change.

The National Council of Supervisors of Mathematics (NCSM) has suggested that the 'twelve essential components of mathematics for the twenty-first century' are: *problem solving, communicating mathematical ideas, mathematical reasoning, applying mathematics to everyday situations, estimation, appropriate computational skills, alertness to the reasonableness of results, algebraic thinking, measurement, geometry, statistics and probability.* I would like to suggest that these are necessary, but not sufficient to make mathematics meaningful and relevant. We will need to contextualise these elements within a broader discussion which involves discussion around issues such as anti-racism, equality of opportunity, raising living standards, justice, critical awareness, cultural respect and developing a global perspective. This calls for a critical pedagogy which is also concerned with empowerment and emancipation from fears of non-understanding and feelings of intellectual inadequacy. Lynn Steen (1990) suggests that we should see mathematics as the language and science of patterns and that these patterns can be explored through the following five concepts:

- *Change*: explore change as it affects real-world issues such as population dynamics.
- *Dimension*: emphasises the need to provide hands-on experience with concepts such as volume, fractals, etc.
- *Quantity*: encourage students to apply math ideas (actions) on things such as census data, inflation trends, sports scores, etc.,
- *Shape*: explores how to create a consciousness of shape theory, for example, maps, photographs, etc.
- *Uncertainty*: how the workings of data and chance affect what happens in our world and how they can help us make plans and decisions.

These may actually suggest a different organising principle for curriculum development at all levels, which could accommodate many of my concerns.

Furthermore, any serious attempt at making mathematics meaningful will have to come to terms with the necessity of abandoning a strict scope-and-sequence approach to curriculum. Curriculum is seen as a programme of activities and not merely as a body of knowledge. Knowledge and skills are constructed from these activities. The curriculum is not determined and fixed prior to teaching, but is something with a problematic status. It is therefore more than likely that students will have "gaps" in the mathematical knowledge in the traditional sense. This is something we will have to accept and use to re-define assessment in mathematics. At the same time we will have to call into question traditional views of education which see the task of the student as internalising and reproducing discrete bits of information. There are assessment practices consistent with this view that knowledge can be broken up into little pieces and can be tested for 'objectivity'. These practices have had a pernicious effect on mathematics education because they tended to devalue, dismiss and separate people. Assessment has sought ways to put individuals in contrast with each other by comparing and rewarding differential individual prowess. We need to develop a view of assessment that does not appeal to this narrow comparative construct, but one that is grounded in the belief that every person is valuable and should be valued. If we are serious about our struggle for social justice and equality, we should actively pursue ways by which we will not assess in order to discriminate, but rather to celebrate the value of each person. In other words, assessment should be illuminatory instead of discriminatory; it should reveal value rather than merely identifying deficiency.

APPENDIX

DISTRIBUTION OF EMPLOYMENT IN MIDDLE AND HIGHER LEVEL OCCUPATIONS 1965-1985 (EXCLUDING 'INDEPENDENT' BANTUSTANS)

	19	65	19	75	19	85
	Male	Female	Male	Female	Male	Female
White						
HLM	175538	57198	306867	95628	389844	141370
MLM	453942	201399	532087	326614	513785	375905
Total	629480	258597	838954	422242	903629	517275
Coloure						
d	7676	10184	16066	17960	21474	33528
HLM	39200	7066	82794	27996	112538	58580
MLM	46876	17250	98860	45956	134012	92108
Total						
Indian						
HLM	5499	1852	10827	4093	17692	8081
MLM	19951	1145	50839	10777	62195	19499
Total	25450	2997	61666	14870	79887	27580
African						
HLM	22724	28338	54468	72137	55392	99574
MLM	89916	7520	189229	19146	274502	67213
Total	112640	35858	243697	91283	329894	166787

HLM: High Level Manpower
MLM: Middle Level Manpower

Source: National Manpower Commission, 1987

EMPLOYMENT IN SELECTED OCCUPATIONS BY SEX AND RACIAL CLASSIFICATION, 1989

Men:	African	Coloured	Indian	White
Engineer	244	227	292	21825
Architect (1)	539	302	186	8712
Natural Scientist	2313	1157	1165	19546
Doctor, Veterinary Surgeon (2)	1714	321	2392	24285
Lawyer	769	85	331	8681
Accountant (3)	1052	592	1650	20588
Nurse	3960	462	124	1170
Educationalist	54063	15085	7713	40803
Secretary	4022	1646	2699	4070
Sales Worker	42413	9845	9967	27781
Caterer	63467	4006	4966	17695
Product Supervisor	14553	6544	3515	43107
Electrician	4210	3139	2099	43790
Metal Worker	10460	8432	3595	82942
Unskilled Worker	3925046	389570	89922	168598

(1) Includes land and quantity surveyor
(2) Includes pharmacist
(3) Includes auditor

Source: HSRC, 1990

EMPLOYMENT IN SELECTED OCCUPATIONS BY SEX AND RACIAL CLASSIFICATION, 1989

Women:	African	Coloured	Indian	White
Engineer	0	4	1	311
Architect (1)	5	3	1	801
Natural Scientist	234	372	207	8179
Doctor, Veterinary Surgeon (2)	447	558	502	7792
Lawyer	26	11	112	1348
Accountant (3)	248	174	420	6463
Nurse	96344	11771	3478	35338
Educationalist	163652	21912	6995	58152
Secretary	6081	9726	6721	114206
Sales Worker	36220	20979	7804	35586
Caterer	92121	26087	1749	21531
Product Supervisor	893	3292	655	1214
Electrician	141	37	16	304
Metal Worker	156	65	22	94
Unskilled Worker	497730	214278	55206	17938

(1) Includes land and quantity surveyor
(2) Includes pharmacist
(3) Includes auditor

Source: HSRC, 1990

REFERENCES

Bell, E.T.: 1945, *The Development of Mathematics* New York, McGraw-Hill.
Bishop, A. J.: 1988, *Mathematical Enculturation: A Cultural Perspective on Mathematics Education* Boston: Kluwer Academic Publishers.
Boyer, C.B.: 1968, *A History of Mathematics* New York, John Wiley & Sons.
Confrey, J.: 1987, *The Constructivist Cycle* Unpublished paper.
Confrey, J.: 1988, Personal communication.
Eves, H.: 1963, *A Survey of Geometry* Volume 1. Boston: Allyn & Bacon, Inc.
Eves, H.: 1969, The History of Geometry. In *Historical Topics for the Mathematics Classroom* Thirty-first Yearbook of the National Council of Teachers of Mathematics. Washington, D.C.
Gattegno, C.: 1965, Mathematics and Imagery *Mathematics Teaching,* 33, 22.
Hogben, L.: 1968, *Mathematics for the Million* London, Merlin Press.
Kline, M.: 1972, *Mathematical Thought from Ancient to Modern Times* New York:Oxford University Press.
Lakatos, I.: 1978, *Mathematics, Science and Epistemology: Philosophical Papers* Vol. 2. J. Worrall and G. Currie (Eds.) Cambridge:Cambridge University Press.
Long, R.L.: 1986, Remarks on the History and Philosophy of Mathematics *American Mathematics Monthly* 93, 609-619.
Meserve, B.E.: 1973, Geometry as a Gateway to Mathematics In A.G. Howson (Ed.). *Developments in Mathematical Education* Cambridge:Cambridge University Press.
Seely-Brown, J., Collins, A. and Duguid, P.: 1989, Situated Cognition and the Culture of Learning *Educational Researcher,*,18(1), 32-42.
Seidenberg, A.: 1960, The Ritual Origin of Geometry *Archive for the History of Exact Science*, 1, 488-527.
Steen, L. A.: 1988, Celebrating Mathematics *American Mathematics Monthly* 95, 414-427.
Steen, L. A. (Ed.): 1990, *On the Shoulders of Giants: New Approaches to Numeracy* Washington, D.C.: National Academy Press.
Wilder, R.L.: 1968, *Evolution of Mathematical Concepts* New York:Wiley.
Zaslavsky, C.: 1973, *Africa Counts* Westport, Connecticut: Lawrence Hill & Company.

John Volmink,
Centre for the Advancement of Science and Mathematics Education,
P.O. Box 17112,
Congella 4013 South Africa

L. BURTON

WHOSE CULTURE INCLUDES MATHEMATICS?[1]

In this chapter I argue that the ample evidence that mathematics is a discipline predominantly attractive to white, male, middle class students can, in part, be explained by the siting, in a socio-cultural context, of its content and the most commonly experienced teaching style. I identify the mathematics encountered through formal education as resonant of a powerful male, eurocentric culture which reifies its own perception of objectivity and reason, in an atmosphere promoting individualism and competition. To address participation rates in mathematics, I claim that we can no longer shelter behind a pretence that the subject is universal, objective and unrelated to the social conditions within which it is developed and practised. Furthermore, the images of the subject itself cannot be separated from the pedagogic style through which it is learnt.

HOW DOES MATHEMATICS FEEL?

> Maths is learning to add, subtract, divide and multiply .... Maths will help us in later life to get good jobs and to work well in our jobs. I don't really like maths, but it is necessary. I think Maths stands for Massive and Torturing Hard Sums. [2]

This is a quote from an eleven-year-old asked to describe mathematics. It tells us quite a lot about that child's view but, I suspect, is not greatly different from the kind of statement that many people would make. It draws attention to three factors which are persistently evoked in discussing mathematics. First, school mathematics, that is the mathematics encountered by the majority of people, is the four rules of arithmetic: "learning to add, subtract, divide and multiply". Second, mathematics is necessary, particularly to the world of work: "it will help us ... to get good jobs". This is more because mathematics is used as a critical filter into employment and professional development than because of its knowledge or skills base as there is evidence that the mathematics which is actually used in the workplace and is more usually generated "on the job" does not directly resemble the formal mathematics taught in the classroom. Both the Bath and the Nottingham study, commissioned for the Cockcroft Report (1982), found that:

> almost all of the mathematics which young people need to use, whatever their job, is included within all the existing ... syllabuses ... Nevertheless, the studies also identified certain important differences between the ways in which mathematics is

S. Lerman (ed.), Cultural Perspectives on the Mathematics Classroom, 69–83.

used in employment and the ways in which the same mathematics is often encountered in the classroom. (para 68/9)

Third, despite acknowledging its necessity, the subject is surrounded by strong negative feelings: "Massive and Torturing Hard Sums". (See, for example, Laurie Buxton, 1981, Dorothy Buerk, 1985). Zelda Isaacson recorded a student's statement:

... like at work ... and they say - what are you studying - and I say, physics, chemistry and maths, - they look at you in absolute horror. (1990: 24)

WHO LEARNS MATHEMATICS?

As teachers of mathematics, we have not been very successful in addressing with learners, what it is about our subject which is engaging. Nor even do we enhance the attainment of most of those who do engage. Furthermore, we continue to teach the subject within the same power paradigm:

Math is eternal, like Latin, and you learn it like you learn any other classic, as an absolute fact. Where did it come from? It just existed, eternally and man (literally, since women weren't doing any of the great math, as far as I knew), well, man had the job of finding that eternal truth and writing it down so we could all know it too. (Buerk, (1985) op. cit. p.61).

As I pointed out on an earlier occasion:

Our secondary schools are geared to achievement in an examination system at which, demonstrably, our pupils' achievements are not even matching the normal curve of distribution ... for many thousands of pupils, ten years of mathematics study has left them with a public record of low achievement. (Burton, 1986, p.5)

And, as Chipman and Brush stated:

Various groups of minority students are more severely under-represented in advanced high school mathematics courses than female students ... black, Hispanic and Native American students were only about half as likely to have taken advanced maths courses as white students whereas Asian American students were about twice as likely to have done so. (1985, p. 323)

So mathematical achievement is differentiated and the achievers are mainly those who are white, male and middle-class (Reyes and Stanic, 1988). Remove one of these categories, for example the last, and the dominance of the other two is maintained, i.e. white males are generally more successful than non-white males and all females. Of course, the phenomenon is not particular to mathematics. Education in general is seen as more appropriate and is more valued when it is perceived as contributing to social, cultural and/or economic objectives. The circularity of this condition thereby ensures that the educational system mirrors societal conditions and reflects its inequalities (see, for example, Apple, 1979, Bowles & Gintis, 1976, Willis, 1977). Jane Kahle pointed out that, in 1982, of approximately 450,000 employed US women scientists and engineers, 85% were white, 7% were black, 6% were Asian and less than 1% were native American. The equivalent figures for males in 1982 were 92% white, 2% black, 4% Asian and less than 1% native Americans (1986, p.71 and p.78). In autobiographical interviews or writings, subjects often speak in terms of families who took it for granted that higher education was appropriate and their decision was which course they would do as opposed to those for whom a decision to attend higher education represented a cultural shift from the norms and expectations of a non-represented culture, into those of the culture which the educational system reflects. Some evidence suggests that women mathematicians are more likely to be only children, or the oldest child in a family and to have attended single-sex secondary schools.

WHAT AFFECTS MATHEMATICAL ACHIEVEMENT?

Laurie Reyes and George Stanic drew attention to five different factors to explain differential achievement in mathematics. They are societal influences, school mathematics curricula, teacher attitudes, student attitudes and achievement-related behaviours, and classroom processes. They pointed out that :

> although we know the five factors are important, we know little about the causal connections among the factors; almost all the extant research is correlational. (1986, p.40)

It seems that in order to account for the differential willingness of certain groups to engage with mathematics and their varying competency when so doing, we must examine how mathematics is perceived and the messages conveyed about it, in and out of school, the siting of these messages in a general context of power and the impact of all of these factors on the experiences and attitudes of learners. For me, this central issue of mathematics learning is exposed by focussing on the *mathematics* and the *learning* and putting these together as a unified experience. I am suggesting that the formal, abstract non-personal presentation of mathematics in school distorts the subject qua mathematics as well as the pupils' experiences and does not address the culture with which they identify and through which they learn.

> The mathematics which is taught is presented as if it is value-free. Because it is dehumanised, depersonalised and (of course) decontextualised, it has been felt

necessary to remove all references to values and other cultural associations in order presumably for the mathematics to retain its 'purity'. (Bishop, 1991, p.13)

I wish to pursue this by raising some questions about the relationship between the perception and experience of mathematics, particularly of those who are still in the early days of their mathematical journey. Throughout, my perspective is one of equal opportunities - the challenge is to reflect on what makes mathematics education accessible and to whom. The questioning mode draws attention to the complex, and problematic, nature of the issue with which we are engaged.

First question: Can a view of *mathematics* as objective, rigorous and convergent be substantiated?

In reading literature in the fields of mathematics education and mathematics, one is struck by the contrast between those who present results, outcomes, 'mathematical products', and those for whom the process of mathematical enquiry is integral to what is being processed. This so-called product/process debate has, to some extent, obscured the distinction between the satisfaction and pleasure brought by mathematical activity and the particular outcomes that it might ultimately accomplish.

For the viewpoint of a user, it is possible, and sometimes even convenient, to identify mathematics itself with its axiomatic presentation in textbooks. From the viewpoint of the producer, the axiomatic presentation is secondary. It is only a refinement that is provided after the primary work, the process of mathematical discovery, has been carried out. (Davis & Hersh, 1983, p.343).

This process is personal, intuitive and culturally dependent; it is informal and creative. Although the outcome of the process is rigorously and logically argued, the journey to this end has usually involved struggle, blockages, and failures, evidence of which, in its public presentation, has been carefully removed. While not wishing in any way to reduce the value of an elegant and rigorously argued piece of mathematics, why should such a high value be assigned to the product in comparison with the mathematical thinking which has produced it? Undervaluing the processes of mathematical thought, also undervalues the mathematician who has engaged in that thought, leaving us with a body of thought which is disembodied. This de-personalisation changes the character of the process and simultaneously reinforces a social view of mathematics as "belonging" only to a very particular set of people, a closed club who subscribe to the same rules and speak the same language.

Mathematics must now be understood as a kind of cultural knowledge, which all cultures generate but which need not necessarily 'look' the same from one cultural group to another. Just as all human cultures generate language, religious beliefs, rituals, food-producing techniques, etc., so it seems do all human cultures generate mathematics. (Bishop, 1988, p.180)

Arguing that mathematics is abstract, objective, and independent of social, cultural and political conditions has left the members of the mathematics club in an elitist and privileged position from which they can attempt to control both the product and those who are admitted to their circle. It leaves the mathematics inaccessible and apparently inappropriate to the majority outside the circle.

Despite rebuttals from those mathematicians who describe their world in terms of its aesthetics, its personal-ness, its culture-boundedness, mathematics continues to be experienced by the majority, especially those still in formal education, as "dry as dust, as exciting as a telephone book, as remote as the laws of infangthief of fifteenth century Scotland" (Ibid, p.169). Nonetheless, or perhaps because of this removal of personal relatedness, mathematical power is not only acknowledged as touching most human concerns but is also seen as doing so in a depersonalised, i.e. objective, abstract and rational, manner. The apparent close links between mathematical, scientific and technological advances (i.e. products) and economic development, have helped to reinforce a reified view of these products in connecting them to the socio-cultural experience which they help to promote.

But, over time even in eurocentric mathematics, there have been shifts in the way in which mathematics has been conceptualised which are largely unacknowledged in its social image or in how it is encountered within education. Mathematics is no longer viewed as a complete system, there has been a shift from certainty to conjecture and from absolutism to relativity. The Platonists in ancient Greece believed that mathematical statements represented eternal truths which were complete descriptions in both absolute terms and in terms of certainty. This belief in mathematical completeness is well represented by Euclid whose system of thought was seen as a complete description of physical space after which no other was deemed necessary or required. The axioms on which the system was constructed were not seen as problematic but as self-evident foundations on which the system could rest with confidence. After the development of non-Euclidean geometries post 1800, a system of geometrical thought was seen to be the result of a particular set of conjectures. Changing the perspective leads to a new set of conjectures which result in a different mathematical model being developed. In the present century, absolutism has succumbed to Einsteinian relativity and Godel dislodged completeness. Proofs are now seen to be demonstrations which take place within fixed systems of propositions and are, consequently, relative to those systems. Objectivity is 'at best' seen to be relative, or abandoned altogether. In 1972, Kline wrote:

Mathematics is not a structure of steel resting on the bedrock of objective reality but gossamer floating with other speculations in the partially explored regions of the human mind. (p.481)

Clearly, there are mathematicians whose images of mathematics are not dominated by notions of objectivity, rigour and convergence. However, the existence of descriptors of mathematics such as these represents one means by which the cultural experience of mathematics can be constrained. Sandra Harding proposed an alternative:

We could look at ... developments in mathematics simply as the onward and upward march of truth in the service of intellectual progress. But to do so hides the social imagery within which numbers and other mathematical notions have been conceptualized, and the very interesting processes of social negotiation through which one cultural image for thinking about mathematical concepts comes to replace another. (Harding, 1986, p.50/1).

For the mathematics educator, therefore, how to broaden not only physical access to mathematics, but more crucially psychological access, becomes a central issue. The social image mystifying mathematics which many of those within the discipline continue to purvey is cyclically transmitted to each next generation of learners through the educative process, not only destroying potential mathematical students but also having a ripple effect on attitudes and performance of those who remain outside of the discipline. The process also ensures that the induction of new mathematicians promotes conformity to this stereotyped, if arcane, image, not least through the teaching styles experienced in university mathematics departments.

In 1976, Lakatos laid down a "long overdue challenge" to the "image of authoritative, infallible, irrefutable mathematics". (p.5) If we, as mathematics educators, accept that mathematics does not have to be experienced as objective, rigorous and convergent, despite persistent presentation of mathematical products reinforcing that view, how can we encourage this socially necessary re-perception of our discipline?

Second question: does a *mathematical pedagogy* based on a transmission model of learning reflect a view of mathematics as objective, rigorous and convergent?

In the famous dialogue between Socrates and the slave, so often cited as an example of good teaching, Socrates was at pains to demonstrate that there was no such thing as teaching, only recollection by the learner of truths already known; the teacher helped in the uncovering process by engaging the learner in dialogue through which the mathematics was elicited. Socrates' role in this was central. Without him, the slave would have been unable to produce the mathematical truths that, it was claimed, he possessed. So as a learner, the slave was dependent upon Socrates. If we are honest, does this not continue as the practice used in many classrooms today? A good teacher "helps" a pupil by judicious questioning and encouragement so that the required mathematical knowledge is recognised and reclaimed. The output of this dialogue is the formal mathematics of the school curriculum, the content of which is validated in the style of the expert teacher. The model of teaching reflects an image which can be represented as the unlayering of an onion, the uncovering process already described. Some teachers use another strategy often described as the filling of an empty vessel to transfer knowledge. Again, the gaining of the knowledge or skill by the learner is dependent upon the expertness of the teacher who assumes a learning void awaiting their action. There are teachers who use both, the first to help the unsuccessful pupil recapture taught knowledge and the second to transfer new knowledge or skill. Pervasive is the assumption that learning is dependent upon transfer from the expert to the naive learner. Consistent with the model is a heavy emphasis on rightness, both of solution and of method. Also consistent is a defined syllabus which is assessed by examination of its content. The formal teaching environment highlights single methods and single solutions. It stresses correctness and product.

However, in his address to the 2nd International Congress on Mathematics Education Rene Thom said:

> The real problem which confronts mathematics teaching is not that of rigour but the problem of the development of 'meaning' ... in practice, a mathematician's thought is never a formalised one ... the only formal processes in mathematics are those of numerical and algebraic computation ... (but) ... even in a situation which is entirely concerned with calculation, the steps of the calculation must be chosen from a very large number of possibilities. And one's choice is guided only by the intuitive interpretation of the quantities involved. Thus the emphasis placed by modernists on axiomatics is not only a pedagogical aberration ... but also a truly mathematical one. (1972, p.202)

A major issue here for the teacher of mathematics is how to ensure that rigour is seen, by the learner, to be an outcome of meaning not a replacement for it. And focussing on meaning rather than product (which is the way rigour is interpreted in many classrooms), requires a different pedagogic style, and consequently, different classroom structures. Elsewhere (see Mason, Burton & Stacey, 1982), it has been suggested that rigour is a response to the need to convince and that one convinces different audiences with different degrees of rigour.

The conception of mathematics transmitted through a formal educational curriculum appears, therefore, to be misguided on two counts, that of the discipline itself, as well as that of the pedagogy. But the consequences are considerable. By emphasising correctness and product, mathematics itself is reified, and the teacher of mathematics is acknowledged as an expert in a difficult domain. However, the ability to perform the mathematical rituals is closely linked to economic and social power and this is reinforced by the way in which mathematics "services" such subjects as science and technology. Restricting access, either deliberately or "incidentally", maintains the elitism and the mysticism. This is one explanation for the frequently reported intellectually "possessive" behaviour of teachers. For example:

I do remember him (the maths. teacher) saying that girls had no place in his field (Isaacson, op, cit p. 23)

Teachers of mathematics were often seen as a stereotyped group exhibiting personality characteristics that were distasteful to the interviewees ... They are 'arrogant' and 'assume knowledge in others', or they 'belittle inability' and are 'dogmatic' and 'lack flair - both social and professional'. (Quilter & Harper, 1988, p. 125)

In recent years there has been growing evidence of the inadvisability of perceiving students as empty vessels and as dependent learners. It has, for example, been conjectured that one reason why the mathematical performance of females is not sustained through the secondary schooling years is because of compliance and consequent dependency on the authority of the teacher, a behavioural pattern which is inadequate to meet increasing mathematical complexity nor to generate a necessary breadth of style and approach to the discipline. (See Scott-Hodgetts, 1986). Quilter and Harper's sample of 15 well-educated adults with a "self-confessed negative attitude to mathematics, and who have little or no confidence in their mathematical abilities" (1988, op.cit. p. 124) far from citing conceptual or intellectual reasons, stressed teaching methodology, its incongruence with learning style and lack of relevance for their failure to achieve.

Although there is widespread acknowledgement among the group that the introduction of 'algebra' certainly does present a major obstacle, the reason for the pupils' difficulty is explained not in terms of the conceptual complexity of the subject-matter, but in terms of its apparent irrelevance and/or the teacher's inability to present it in a coherent, meaningful way. (Ibid, p. 127)

If we shift our pedagogical stance away from the didactic, delivery model, there are consequent effects on the subject and the learners. Evidence is accumulating as to the close interaction between learner autonomy and confidence leading to engagement and success. Some of this evidence comes from those applying techniques for writing-to-learn-mathematics. (See, for example, Countryman, 1992). Other evidence is offered by those exploring error analysis. (See, for example, Borasi, 1987) A considerable body of evidence, internationally, is accruing through a focus on alternative styles of assessment. (See, for example, Stephens and Izard, 1992 and Burton, 1994) What are the most effective means of encouraging teachers to make shifts intellectually in re-perceiving mathematics, and pedagogically in classroom practice? In the psychologically potentially stressful situation of the classroom, how can teachers be persuaded to abandon the comparative security of their knowledge base in favour of learner creativity, independence and autonomy all of which could be seen as increasing the psychological load?

Third question: What pedagogical stance is appropriate to a view of mathematics based on incompleteness, conjecture and relativity?

A different metaphor, that of construction and exploration, has been proposed to challenge the assumptions of transfer and uncovering (See, Davis, Maher and Noddings, 1990). In constructing and exploring, the emphasis shifts from product to process, from single solution to alternatives. Students are acknowledged as lively and profound thinkers, engaged upon trying to make sense of the new in the context of what is already understood. The teacher's role changes to one of resource provider, rather than explainer. The students are placed in a group situation with a question which they find challenging. The resultant cognitive conflict provokes them to work together towards a resolution which makes personal, and inter-personal, sense.

For teachers this is not unproblematic. As the pace, content and outcomes of an activity shift into the hands of the pupils, the "normal" expectations of teacher and pupil roles change substantially. It becomes clear that the teacher's accessibility to everyone in any one lesson is limited by her/his need to observe and interact with small groups or individuals. The time scale of the enterprise has to change. The participants not only assume different roles but also different responsibilities. The teacher has to accommodate to students pursuing their own lines of enquiry, and managing their learning independently. Clarity of transmission becomes less important than how perceptive are the teacher's observations of, and interactions with, pupils' cognitions and how effective are the associated learning plans. Collecting detailed verbal and non-verbal observations makes available to teachers a wide range of information on pupils' understandings and mis-understandings. Comparing such observations with their conventional 'expectations' often leaves teachers astonished both by what learners know, and can do, as well as by what they cannot.

Outstanding questions to be addressed in respect of this pedagogic shift remain. For example, of what does a curriculum consist? What happens to the distribution of power both in relation to the discipline and to the learning? If power resides in the learning community rather than the community of scholars or the society at large, how is this acknowledged, developed, celebrated? When and what does a teacher "tell", does skill learning always sit comfortably within skill using, what professional judgements is it reasonable to expect of a teacher and how do they acquire the knowledge to make these?

Fourth question: What impact does this shift in mathematical pedagogy have on accessibility to mathematics?

I have claimed that the pedagogical processes which are most common in the teaching of mathematics deny the influence of the individual or their social context and present a pretend world of certainty, exactitude and objectivity. I believe that this world is associated with power and control and with images of mathematics and science that are locked into what has been called a "pervasive and fundamental dualism, a tendency to create frameworks for viewing the world in terms of bimodal categories, with attendant evaluations" (Weinreich-Haste, 1986, p.116. See also Glennon, 1979 and MacCormack and Strathern, 1980). The dualism to which Helen Weinreich-Haste draws attention resides in the opposition of reason, conventionally linked to the scientific method, and emotion and the mapping which is made into gender so that males become the exemplar of scientific reasoning, whereas females are identified with emotion and nature. She pointed out that:

> the primary image of scientific endeavour which recurs throughout the modern period is of Nature, as female, being conquered, penetrated and yielding up her mysteries to the male scientist. Note that as well as the obvious sexual symbolism of these images and metaphors, the relationship implies control, domination and the imposition of order and reason. (Ibid, p. 118)

Throughout our social history, power and control have been vested in particular races and particular sexes. By perpetuating a model of teaching which rests upon knowing, and consequently upon 'expertness', we reinforce this hierarchical view. As George Joseph has made clear, by validating a depersonalised, elitist and formal model of mathematics, we ensure that the subject remains aloof from the concerns and interests of most members of society.

> First, there is a general disinclination to locate mathematics in a materialistic base and thus link its development with economic, political and cultural changes. Second, there is a tendency to perceive mathematical pursuits as confined to an elite, a select few who possess the requisite qualities or gifts denied to the vast majority of humanity. This is a view prevalent even today in the classroom and thus determines what is taught and who benefits from learning mathematics. Third, there is a widespread acceptance of the view that mathematical discovery can only follow from a rigorous application of a form of deductive axiomatic logic which is perceived as a unique product of Greek mathematics ... finally, the presentation of mathematical results must conform to the formal and didactic style following the pattern set by the

Greeks over 2,000 years ago ... as a corollary, the validation of new additions to mathematical knowledge can only be undertaken by a small, self-selecting coterie whose control over the acquisition and dissemination of such knowledge through journals has a highly Eurocentric character. (Joseph, 1987, p. 22/3)

I believe that only by moving mathematical pedagogy into the same world of conjecture, relativism and incompleteness that matches the mathematics already so described will we succeed in breaking the image of mathematics as european and male and in creating the kind of classroom in which all students can find excitement, interest and success. It seems to me to be a supreme irony that students disengage from the subject known as school mathematics, which itself is rejected as an adequate representation by many professional mathematicians. Were mathematics to be experienced as a searching, hesitant, intuitive area of study, open to interpretation and challenge, I feel confident that there would be a much greater identification with its style and ideas by pupils from both sexes, different classes and different races. Accumulating evidence (for example, Gerdes, 1985, Schoenfeld, 1989) reinforces the view that a competitive style of learning is destructive for a lot of students who change their view of mathematics, and of their potential to learn the subject, when placed in a climate which encourages them to work together, listen and learn from each other, explore and respect different perspectives.

A CONCLUDING AGENDA

Where are we now? We already have an extensive list of work which is urgently needed in order to establish the inequalities of the current situation. We await evidence from many different societies of the impact of Reyes and Stanic's five factors on differential group performance and the causal connections between them. Sensitive ethnographic studies are necessary to listen and learn from the pupils about their mathematical experiences and expectations. It is always easier to operate a deficit teaching model which assumes that failure to learn is because of learner deficiency. However, there is now abundant evidence that in most cases if children fail to learn, or mis-learn, it is neither deliberate on their part nor a function of inability. More likely, it is because the teacher has not taken into consideration evidence about the learner which could provide both explanations for how or why the mis-learning happened as well as indications as to what kind of experience might provoke a readjustment of that learning. In ignoring this evidence, teachers often over-simplify the material in an attempt to "convey" it. At one and the same time this patronises the learner and fails to affect the status of their knowledge. Eleanor Wilson Orr wrote of:

more and more situations where teachers, understandably discouraged with high failure rates, gradually, and more often than not unconsciously, modify what they do and require. It is painful to face unremitting large numbers of students who do not understand and to be unable to get them to understand; good teachers blame themselves and try again and again. Eventually uncertainty leads to adjustments, bit by bit, to what students *can* do. (1987, p. 202).

In her book *Twice as Less* she revealed how differences between Black English vernacular speech and standard English can impede the learning of mathematics for Black English speakers. She discussed a reliance on what she called learning by pattern. By this she meant the identification of a learned response, often a form of words, with a particular mathematical situation and the reproduction of that response without thinking through its meaning. Some might call this over-generalising:

The students must encounter these verbal tools when they need them - when they are engrossed in trying to figure something out. But for this to take place, the habit of learning by pattern must be broken; the dependence on the replicable as the only guarantee of being correct must be replaced by the habit of thinking. These students must experience their own minds; they must trust their own intelligence. What they learn in mathematics and science must therefore make sense: it must fit their sensory world - what they know directly through their senses, and what they can know indirectly by deliberate derivation from their sensory experience. They must come to know that in mathematics and science there is a reason for everything; they should discover the emptiness of knowing only a pattern, the incompleteness of not seeing a reason. This is important for all students, but for these students it is the way to unlock their natural intelligence. (Ibid, p.205)

Her analysis not only revealed why certain mis-understandings arise and indeed how natural it is that students should get caught in the conflict created by inconsistencies of language usage but she also demonstrated how adopting a supportive but challenging stance can alter the situation for learner and teacher.

Mary chose to define a right angle as an angle whose rotation is one-fourth of a complete rotation, and perpendicular lines as the sides of a right angle, while Robert ... chose to define perpendicular lines as two straight intersecting lines that form equal adjacent angles and a right angle as an angle whose sides are perpendicular. Thus Mary and Robert went about proving a given statement differently, fully aware that one way can be as correct as the other. Furthermore, because proofs were also done orally, students knew that however they worded the steps of their proofs, each step had to be clear to the rest of the class; their reasoning had to make sense; and they had to be ready if challenged to defend their proofs. (Ibid, p.210)

This approach opens the door to the investigation of consistencies and inconsistencies in mathematical language use across many cultures not only as an explanation for differential performance but also as further incentive to change pedagogical expectations and experiences. It also projects an image of a classroom where individuals have the right and obligation to interpret, exemplify, generalise and convince others and to respect and attend to the differences and the similarities thus provoked. This is very much along the lines of what Alan Schoenfeld refers to as "membership in a community and ... the ways that one comes to view the world by being a member." (op. cit. p.86)

Where are we going? We need to take seriously Lakatos' challenge to reject the presentation of mathematics as "an ever-increasing set of eternal, immutable truths" in which

> counterexamples, refutations, criticism cannot possibly enter. An authoritarian air is secured for the subject by beginning with disguised monster-barring and proof-generated definitions and with the fully-fledged theorem, and by suppressing the primitive conjecture, the refutations, and the criticism of the proof. Deductivist style hides the struggle, hides the adventure. The whole story vanishes, the successive tentatives formulations of the theorem in the course of the proof-procedure are doomed to oblivion while the end result is exalted into sacred infallibility. (1976, p.142)

The implications of taking this seriously lead not only to a revision of mathematical "truth" and a re-presentation of mathematical knowledge at every level as subject to review and criticism but also to a reformulation of relationships in the classroom so that the construction of mathematical knowledge is seen by all concerned as responsive to a process of conjecture and critique. To implement an atmosphere which effectively sustains the learning of mathematics in this way demands that teacher and learner respect each other's contributions to the process. At the same time, we need to develop and evaluate different learning programmes which reflect the richness in the classroom and society. We need evidence to support our intuitions that mathematics which is person-related, culturally embedded and meaningful becomes more accessible to a wider constituency of learners. Some evidence is already available but it is fragmentary and itself subject to ethnocentrism in that it frequently does not pass outside the boundaries of the institution or country where it has been gathered.

The child's growing interest and skill in pursuing mathematical enquiry depends upon sustaining her interest in the world around her such that she can formulate questions about it and develop her capacity to attend to, focus on and investigate aspects of that world. This does not require 'mastery' or domination although those are the terms used so often by mathematicians. Far from being a process of closure, as experienced by so many students, it is cyclical where choices dictate new ideas, future questions, further investigations. Such enquiries are not conditioned by power and domination but thrive in an atmosphere of collaboration and connectedness where alternative perceptions, interpretations, methods are part of the richness. An environment which combines intellectual questioning with social interaction might prove to be the most effective for providing a cultural context which is meaningful to the majority of mathematical learners.

NOTES

1. This chapter is an amplification and development of a paper given on the Fifth Day of ICME 6 and subsequently published as 'Mathematics as a Cultural Experience: Whose Experience?' in Keitel, C. et al (Eds) (1989) *Mathematics, Education, and Society*, Paris: UNESCO.
2. Quoted from Burton, L. (1989) 'Images of Mathematics', in Ernest, P. (Ed.), Mathematics Teaching: the State of the Art, Lewes: The Falmer Press, p.182/3.

REFERENCES

Apple, M.: 1979, *Ideology and Curriculum*, London: Routledge & Kegan Paul.

Bishop, A.J.: 1988, 'Mathematics Education in its Cultural Context' in Bishop, A.J. (Ed) *Mathematics Education and Culture*, Dordrecht: Kluwer.

Bishop, A.J.: 1991, *Mathematical Enculturation*, Dordrecht: Kluwer.

Borasi, R.: 1987, 'Exploring Mathematics Through the Analysis of Errors', *For the Learning of Mathematics*, Vol. 7, No.3 (November), pps 2-8.

Bowles, S. and Gintis, H.: 1976, *Schooling in capitalist America*, New York: Basic Books.

Buerk, D.: 1985, 'The Voices of Women Making Meaning in Mathematics', *Journal of Education*, Vol. 167, No.3, pps 59-70.

Burton, L.: 1986, *Girls into Maths Can Go*, London: Holt, Rinehart & Winston.

Burton, L.: 1994, *Who Counts? Assessing Mathematics in Europe*, Stoke-on-Trent: Trentham Books

Buxton, L.: 1981, *Do you Panic about Mathematics*?, London: Heinemann.

Chipman, S. F. and Wilson, D. M.: 1985, 'Understanding Mathematics Course Enrolment and Mathematics Achievement: A Synthesis of the Research', in Chipman, S. F., Brush, L. R. and Wilson, D. M. (Eds.) *Women and Mathematics: Balancing the Equation*, Hillsdale, NJ and London: Lawrence Erlbaum Association.

Cockcroft, W. H. (Chair): 1982, *Mathematics Counts*, Report of the Committee of Inquiry into the Teaching of Mathematics in Schools. London: HMSO.

Countryman, J.: 1992, *Writing to Learn Mathematics*, Portsmouth, NH: Heinemann.

Davis, P. J. & Hersh, R.: 1983, *The Mathematical Experience*, Harmondsworth: Penguin.

Davis, R.B., Maher, C.A. and Noddings, N. (Eds.): 1990, 'Constructivist Views on the Teaching and Learning of Mathematics', *Journal for Research in Mathematics Education*, Monograph No. 4, Reston, Va.: N.C.T.M.

Gerdes, P.: 1985, 'Conditions and strategies for emancipatory mathematics education', *For the Learning of Mathematics*, 5, 1.

Glennon, L. M.: 1979, *Women and Dualism*, London: Longman.
Harding, S.: 1986, *The Science Question in Feminism*, Milton Keynes: O.U. Press.
Isaacson, Z.,: 1990, 'They look at you in absolute horror: Women writing and talking about Mathematics', in Burton, L. (Ed.), *Gender and Mathematics: An International Perspective*, London: Cassell.
Joseph, G. G.: 1987, 'Foundations of Eurocentrism in Mathematics' *Race & Class* 28, 3, pp. 13-28.
Kahle, J. B.: 1986, 'From One Minority to Another' in Harding J. (Ed.) *Perspectives on Gender and Science*, Lewes: The Falmer Press.
Kline, M.: 1972, *Mathematics and Western Culture*, Harmondsworth: Penguin.
Lakatos, I.: 1976, *Proofs and Refutations*, Cambridge: Cambridge U.P.
Mason, J., Burton, L. and Stacey, K.,: 1982, *Thinking Mathematically* Wokingham:Addison Wesley.
MacCormack, C. and Strathern, M.: 1980, *Nature, Culture and Gender*, Cambridge: Cambridge University Press.
Orr, E. W.: 1987, *Twice as Less*, New York: W. W. Norton.
Quilter D. and Harper E.: 1988, 'Why we didn't like mathematics, and why we can't do it', *Educational Research* Vol. 30, No. 2, June, pps. 121-134.
Reyes, L. and Stanic, G.: 1988, 'Race, Sex, Socioeconomic Status and Mathematics' *Journal for Research in Mathematics Education*, Vol. 19, No. 1, pp. 26-43.
Schoenfeld, A.: 1989, 'Peer Interaction, Problem-Solving and Cognition', *International Journal of Educational Research*, Vol. 13, No. 1, pp. 71-87.
Scott-Hodgetts, R.: 1986, 'Girls and Mathematics: The Negative Implications of Success' in Burton, L. (Ed.) *Girls Into Maths Can Go*, London: Cassell.
Stephens, M. & Izard, J. (Eds.): 1992, *Reshaping Assessment Practices: Assessment in the Mathematical Sciences under Challenge*, Hawthorn, Vic.:Australian Council for Educational Research.
Thom, R.: 1973, 'Modern Mathematics: Does it Exist?' in Howson, G. (Ed.) *Development in Mathematics Education*, Cambridge:Cambridge University Press.
Weinreich-Haste, H.: 1986, 'Brother Sun, Sister Moon: Does Rationality Overcome a Dualistic World View?' in Harding, J. (Ed.) *Perspectives on Gender and Science*, Lewes:The Falmer Press.
Willis, P. E.: 1977, *Learning to Labour: How working class kids get working class jobs* Farnborough:Saxon.

Leone Burton,
School of Education,
University of Birmingham,
Edgbaston,
Birmingham B15 2TT U.K.

R. PINXTEN

ANTHROPOLOGY IN THE MATHEMATICS CLASSROOM?

INTRODUCTION

Anthropology used to be the study of exotic, or at the least non-western cultures. Although its range has been expanded over the decades to include the study of western subjects and that of intercultural interaction and communication (be it in the West or elsewhere), mathematics and mathematics education is still rarely connected to anthropology. So what can anthropology offer to the mathematics teachers? In my opinion, insights from at least four different foci of anthropological research are relevant for the mathematics teacher in general, whichever cultural group she is teaching: types of learning; cognitive contents; language structure and institutional aspects of teaching. Some of these may be closely interrelated in a particular perspective (e.g., language structure and cognitive contents for the linguistic relativist - see below), but analytically each can be distinguished from the next one. Each one of these imply both theoretical and practical knowledge. However, before I start on this journey, I have to state my position in the debate on the nature of the subject matter of teaching, that is on the nature of mathematics.

A Note About The Nature Of Mathematics

In recent publications the nature of mathematics teaching has been discussed explicitly in the context of a particular understanding about the nature of mathematics. I will be content to mention only a few of the most recent contributors to this fascinating discussion. Ernest (1991) devotes roughly a third of his book to an overview of the different positions which are still defended vis-à-vis the status of mathematical knowledge. He criticizes and rejects the three "absolutist" views (formalism, logicism and constructivism) which are all versions of deductivism, and opts for a fallibilist perspective that founds the relative dependability or safety of mathematical knowledge both in empirical facts and in faith, but not in a priori and hence indubitable, givens. He thus carries further and deepens the studies of a growing group of scholars who characterize mathematical knowledge as 'contextualized' knowledge, that is knowledge which draws meaning, relevance and truth from its cultural, social and/or historical context. The distinction between 'absolutists' and 'empiricists' identifies the two main perspectives in the debate. 'Empiricists' such as Ernest (but also Wilder, Bishop, Kitcher and a

S. Lerman (ed.), Cultural Perspectives on the Mathematics Classroom, 85–97.

few others) show that the deductive type of reasoning has to start from intuitive givens or presuppositions about axioms, concepts and rules of inference. These intuitions and presuppositions are drawn from the reasoner's experiential knowledge. Hence, the apriorists in the discussion on the nature of mathematical knowledge underestimate or deny the experiential basis in their thinking. Bishop takes a very intriguing and indeed radical position in this debate. In his recent work (e.g., 1988a and 1988b) he defines mathematical education as a particular social (and cultural) process. This view correlates his vision that mathematics is "essentially a ' symbolic technology'.." (1988a p.18). That is to say, it is primarily a way of using signs, techniques, procedures and the like in practice.

He then goes on and searches for instances of this technology in different cultures: he suggests, on the basis of a host of empirical studies, that six 'environmental activities' form the actual basis of what is recognized as mathematical thinking. These activities can be found, in varying forms and to a different extent, in all cultures. They are: counting, locating (spatial acting, seeing and thinking), measuring, designing, playing and explaining (Bishop 1988a and 1988b). This empirical foundation of mathematical thinking in the cultural context of the knower has two important benefits, as I see it: first of all it allows the researcher to investigate in an unambiguous way the relationship between the processes of knowing and learning (be it for mathematical or other types of knowledge) on the one hand and the cultural and social context of the knower on the other hand. Bishop warns: "In trying to understand mathematics as a cultural phenomenon we must take care not to decontextualize the ideas too rapidly" (1988a p.33). In the second place it allows for a comparative or cross-cultural analysis of this type of knowing. In doing so it opens the road towards a profoundly emancipatory view on mathematics education. Bishop repeatedly claims that the old attitude of teaching a fixed and somewhat esoteric body of mathematical knowledge to a group (a select group of intending mathematicians as some profess, or everybody in his or her own cultural setting as others propose) can now be abandoned for the general humanistic program of a genuine mathematical education for all. The latter refers to the intention to educate all layers of society and all different cultures about, through and in mathematical knowledge. In doing so the culture-specific activities and metaphors of the learner will be the basis on which to gradually build a full grasp of mathematical knowledge in any random group.

If the historical picture of Davis & Hersh (1981 but even more so 1986) is correct that the economically and politically dominant Western society is more and more framed in and dependent on mathematical skills, and that success and survival in the world of the future will be dependent to a large extent on one's ability to be 'literate' in mathematics, then Bishop's approach implies a decidedly ideological choice. Indeed, if mathematical ability is growing into a possibly discriminatory criterion against a person or a group (the haves and have-nots are redefined along this line - I will skip the literature on this point), then Bishop's suggestion offers an emancipatory program against a growing discrimination.

This short introduction suffices. It should be clear that I take sides with the 'empiricists' in the debate on the nature of mathematics. The implications for mathematical education have already been drawn to some extent in the works by Bishop and Ernest mentioned above. However, I think a lot more can and should be said about the 'cultural context' they refer to. This will be the substance of my contribution to the discussion.

LEARNING AS A CULTURAL PHENOMENON

It is trivial to say that mathematics education is primarily a learning process. However, the understanding of what this learning process actually is, is far from obvious. As an anthropologist it strikes me that the discussion on the nature of learning is narrowed down to a small set of alternatives which all bear on the same basic paradigm. The alternatives can be phrased along differences in ideological positions (Ernest 1991, chapter 7): conservatives train pupils in particular unalterable skills; pragmatists adapt the curriculum in view of professional needs; old humanists train an elite to fully develop an inherited cast of mind; progressives make use of local cultural givens to transfer the skills; and social constructivists approach mathematics critically in order to further the development of a general attitude of democratic citizenship. These distinctions are certainly important to make, but they focus on the ends of mathematical knowledge, rather than on the structural and process aspects of learning itself. Other authors focus on the different types of teaching principles: imitation, rote learning, stimulus-response conditioning, and so on (e.g., Davis, Carraher, and others in the ICME section published by Keitel et al. 1989). Basically, these teaching principles express one or the other form of behaviourism or cognitivism. The reason is simple: behaviourism and cognitivism are the only existent learning theories in the West as of today (see Cole & Cole 1989). Within this pair a choice can be made between one or the other learning and teaching theory, and modifications on each can be worked out. However, this couple is decidedly Western. In other words, the difference expresses nuances within the Western paradigm. Hence, it is unwarranted to attribute a universal status to this type of learning, even implicitly. My suggestion is to be aware of other types of learning in other cultures and to indicate their relevance for teaching mathematics in these cultures.

The work by Cole et al. (1971) on the introduction of New Math with the Kpelle in Liberia is widely known. Among other things it showed that the use of measuring and counting procedures is linked for the Kpelle with particular cultural tasks and specific contexts (like the market, manufacturing of cloths, and so on). In other words, the Kpelle are not inclined to use their mathematical capabilities out of particular contexts. They also learn this type of knowledge in the appropriate situations. Cole et al. came to understand all this after they tried to conduct the typical psychological experiments and tests with the Kpelle, where knowledge and skills are presupposed to be used and applied in decontextualized formats (like a school situation, or a test situation); the knowledge of the apprentice is linked to persons, situations and tasks and there is no motivation to use it "in the void", that is in a decontextualized way. This finding is generalized in the work of a member of the same research group, referring to mathematical and other kinds of knowledge (Lave 1988). What is at stake here, I think, is a different kind of learning altogether. Let me try to illustrate this by means of examples from one of the two foreign cultures I know more about.

The Navajo are North American Indians, living in the Southwest of the USA. They have a long standing tradition of learning through storytelling. Roughly, the principal features of Navajo learning can be contrasted in a general way to those of the Western school-type of learning. The following dichotomy is a simple way to represent differences between both traditions, but it cannot be held to correspond to simple point-by-point contrasts. I claim that Navajo learning shows:

1) more emphasis on qualitative ordering and aesthetic aspects and less on quantification and universal statements;

2) more stress on orthopraxy (to behave properly, appropriately, and so on) and less on orthodoxy (to share the same contents as the other members of the group). Knowledge is contextual, person- and problem-bound.

A concrete effect of these two characteristics (1 and 2) is that the Navajo child observes very carefully and with extreme precision, but nearly never asks questions. In the classroom this leads to awkward misunderstandings, because the school system appreciates questions as a sign of curiosity and active involvement in the programme on the part of the child, and it encourages trial and error action against the silent observation that is 'typically Navajo'.

3) more dependence on the persons involved in knowledge transfer, and much less room for a curriculum format and hence for a universal status of knowledge.

4) more awareness of the negotiation aspects of each learning situation and less respect for the institutional authority of a teacher. What can be known and learned depends heavily on how the personal characteristics (such as knowledgeability, age, trust, etc.) of all involved and the situational relevance of the item will be agreed upon in each learning process. Neither the status of the teacher nor that of the knowledge to be learned are fixed once and for all or across contexts for the Navajo. The school system does the opposite for the Westerner; the teacher is the one who knows, and the curriculum material is the kind of knowledge that should be taught, regardless of the opinion of the pupils.

Because of the contrastive presentation of both traditions of learning (from the learner's point of view) and teaching the picture that emerges may be a bit simplistic (but see Pinxten & Farrer 1990). Still, I hope to have conveyed at least the general point that learning styles are very different from one culture to another. Moreover, one of the features of contrast that appears to stand out in all the studies mentioned so far is that between a view on learning as depersonalized and decontextualized transfer of information (as was and is the case in the most extreme version of school learning) and an understanding of learning as a person- and situation-dependent transfer of appropriate knowledge. In my understanding the disregard of this difference and the so-called culture-free implementation of the school format of learning in other cultures is one of the major elements which lead to vast difficulties for children from other cultures in Western style schools.

In the mathematics classroom it is perfectly possible to cope with differences in learning styles, provided one is willing to alter or altogether disregard the curriculum as the source of knowledge and to found one's teaching theory on the 'other' learning style of the pupils. Some of this is worked out for working class people of Western cultural origin by Frankenstein (1989), taking the world of experience of these people as the source of mathematical problematization. Another example is the alternative curriculum for geometry teaching which I worked out for Navajo children. Problems, concepts, criteria of relevance and terms are taken from the preschool knowledge in the experiential world of the Navajo. These elements are rediscovered by the child and hence made more conscious in the mind and the language of the Navajo by presenting them as a context for learning in the classroom (Pinxten et al. 1987).

KNOWLEDGE CONTENTS ARE CULTURE SPECIFIC

Davis & Hersh (1986) described how the modern world in the West is full of mathematical applications and concepts, to the extent that even the mathematician is unable to master everything nowadays. In a subsequent paper Davis (1989) redefines the aims of mathematics education in the West by stressing the importance of applied rather than pure mathematics for the common pupil. He draws the parallel with cars; although only a few can actually make a car, all of us can drive. Similarly, all of us should learn to "drive" mathematically. In practice, he suggests the following general lines on which to base mathematical education:

(a) Identify and describe the mathematical beliefs, constructions, practices that are now in place. Where and how is mathematics employed in real life?
(b) Describe the mathematical beliefs, constructs, and practices that have been identified by the community. What are justifiable and what are unjustifiable? What are the modes of justification?

(c) Describe the social dimensions of mathematical practice. What constitutes a knowledge community? What does the community of mathematicians think are the best examples that the past has to offer? (Davis 1989 p.27)

The kinds of topics that are proposed here by a mathematician are precisely those that I indicate in the subtitle as 'contents of knowledge'. Davis looks at the use and spread of themes and concepts in the broader cultural context of the Westerner, and he urges us to consult the past of this culture in order to find the most suitable examples for teaching. On my proposal similar questions should be at the basis of the mathematics education of every group. In this regard it is relevant to be aware that we are beginning to gather substantial knowledge about spatial notions in other cultures (Kearney 1983, Lewis 1972, Pinxten et al. 1983). This ethnographic knowledge indicates that, within certain constraints, space is experienced and processed in a wide variety of ways. To make it quite clear, I will cite a detailed example: for the Navajo Indian 'movement' is a constituent of nearly all spatial distinctions, including such 'static' or structural notions (at least in the Western view) as angles or volumes. The difference is not necessarily in terms of the units or entities discerned in reality, but rather in the conceptual buildup: the category "to go" is the Navajo correlate to the Indo-European category "to be" (Pinxten et al. 1983). At the same time, we now know different counting systems across the world. Among a growing list of studies I can only mention a few: different numerical units and diverging ways of adding and subtracting in Africa have been detailed by Zaslavsky (1979), whereas Dawe in Australia (1988) and Lancy (1983) in Papua New Guinea were able to detail different counting systems (using the human body and other 'carriers' as devices). Information on Native American geometry and number systems can be added to this thesaurus of living and practical counting procedures (Closs 1986, Pinxten et al. 1987). The older but still very eminent source for Chinese mathematics and geomancy remains Needham (1965, esp. Vol. 2). A first attempt by an anthropologist of an overview and synthesis of the field, at least with regard to numbers, is to be found in Crump (1990). More regionally restricted overviews are numerous; a better example spanning the whole world (with a strong emphasis on the West and the ancient Middle East), is Ifrah (1985). The tremendous, but very diverse, information on geomancy systems and other divination practices should be consulted as well to find more 'applied mathematics' in the cultures around the world. To mathematicians all this may appear to be 'primitive' and 'on the way out'. Anthropologists keep reminding their colleagues that these and other cultural features are very tenacious and continue to be used and learned, apart from and parallel to school knowledge. The growing interest in 'ethnomathematics' (i.e., mathematical notions and skills in a culture) is a further indication that the different contents can and will profitably be used as a basis of inspiration for mathematics teaching (D'Ambrosio 1985 and 1987).

LANGUAGE STRUCTURE

The old hypotheses of linguistic relativism (e.g., in its best known form: Whorf's hypothesis in Carroll 1956) had it that the structure of a language (co)determines the thought of the community of speakers. The language structure (e.g., its syntax) is stable and nearly invariant over generations; therefore it structures thinking and behavior. The general hypothesis has been formulated in a variety of ways. The strongest version has been abandoned, but the weaker version (speaking of one-or two-directional co-determination or strong influence from language structure on thought) seems to be generally accepted by now. The hypothesis is relevant in the present context, because mathematical knowledge is a kind of knowledge expressed in a language. Hence the relationship between both and the culture specific nature of both is highlighted. Indeed, the commonly used mathematical language has a structure that is decidedly that of the European languages; it distinguishes clearly between things (classes, categories, sets, etc.) and operations on things (sum, division, etc.). The 'view on the world' which is implicit in mathematical knowledge corresponds to a large extent to that of the ontology of the Westerner; reality consists of things (subjects, objects, situations, etc.) and actions or events vis-à-vis these things (movement, change, etc.). At a more specific and conscious level, a generic atomism is projected as the basic order of reality for the Westerner: things can be understood as wholes which consist of constituent parts, and as parts in an englobing whole. One can know the world by dividing it in smaller parts and study those, and by combining knowledge about parts one can understand the greater whole. Not only the verb-noun dichotomy, but also this part-whole reasoning is essential in mathematics too; a set has elements, a geometric figure has parts (center, surface, etc.) and counting, algebra and geometry are taught by means of part-whole metaphors. (Whether or not the language structure actually determines this way of reasoning is not important. For the present argument it suffices to see that this way of dealing with reality is at least matched in the structure of the European languages.)

Against the background of all this, it is very striking for an anthropologist to come across languages with a very different structure. Chinese, and Athapaskan languages are only two of a series. For none of these the part-whole distinction is either as prominent as or really the same as in the European languages. To put it more concretely, a Yin-Yang model of the universe does not describe the world merely in terms of Yin and of Yang (classifying one 'thing' as Yin and another as Yang, for example). It does not work that way: rather, all is in flux, and the world is more like an organism than like an atomistic structure (Needham, op cit). In the case of the Athapaskans (Apache, Navajo and so on) the world is seen as movement, dynamic forces and the like, rather than things or structures. To put it in

Whorfian terms: a Navajo does not say 'the sky is blue', but rather 'the blueing is taking place' (e.g., Witherspoon 1977). The overwhelming dominance of verbal forms in the Athapaskan languages points to this way of dealing with the world. The part-whole distinction is really marginal in the vernacular (Pinxten et al. 1983 p.65-68). A more general conclusion has been formulated by a distinguished linguistic anthropologist who evaluates the claim of genetic (and hence universal) status of the deep structures in the transformational generative grammar linguistics of Chomsky and his followers:

> The most damning evidence against the transformationalist argument ... is that some of the so-called fundamental syntactic devices of deep structure are not available in all languages, and even when available are not equally developed or equally favored. (Tyler, 1978 p. 69)

When mathematics is introduced to a child with a background in such a culture, it is clear that misunderstandings and even complete puzzlement will occur. Any insightful learning is blocked, according to my view, because the 'world view' implicit in mathematical knowledge does not correspond with the way the world is handled in the preschool knowledge of the child. The only way out of this frustrating situation I can suggest, is to conceive of mathematics education as 'enculturation' (Bishop 1988). That is to say, the mathematical activities in the culture of the child should be seen as the principle entries to (school) mathematics. These activities function as means to handle the world in each culture, hence they should be the cornerstones of mathematical education in the school. A net result of such a position is, of course, that curricula should be specific to the cultural group, if they should be at all. Again, a practical example from another culture I know a little bit better will clarify what I mean by this.

In a curriculum for geometry teaching with Turkish immigrant children in Belgium I propose to have the children explore their neighborhood as a first project context. The course is meant for children of 7 to 9 years old. The neighborhood is the micro-world which is 'Turkish' (with the character of an open ghetto, one might say) and it is much better known than the surrounding 'Belgian' world. Paths, distances, places and so on structure this micro-world. The first step in the geometry course is to have the children re-explore it consciously. In the classroom all terms and categories used are then collected and written down (in Turkish, and in a second period in Flemish- the language of the dominant culture), the neighborhood is reconstructed in a scale model by means of waste material, after which it is graphically represented. In the whole process the vernacular and the cognitive categories of the children are the only "knowledge" the teacher has them work with in the classroom. Gradually, in their own language and in their own cognitive frame the children will 'reinvent' the geometric notions they need in order to rebuild, draw and measure aspects of the neighborhood.

The importance of using the language resides in its status of carrier of meaning; particular notions (e.g., for corner, front and back, and the like) conflict or are incongruent with their correlates in Flemish. Having the children 'reinvent' geometry on their own allows them consciously to discover the differences and to adapt their imprecise or even 'mistaken' notions where and when their insight induces them to. The differential functionality of natural language and of geometry will thus be insightfully learned by the children, without frustration out of dominance and lack of understanding (Pinxten mimeo).

INSTITUTIONAL ASPECTS

I can be rather brief about this last dimension. Mathematics education always takes place in an institutional context. The context that is dominant, and for which both teacher training and curriculum development programmes have been set up over the years, is the school context. However, as has become quite clear in a variety of studies referred to so far, the school is not the only and maybe even not the most prominent institutional setting for the learning of mathematical skills and notions. The beautiful work by Cole et al. (1971) showed that Kpelle learn and use mathematical notions in their professional life (e.g., on the market place), whereas Lave (1988 and elsewhere) demonstrates how geometric notions function in the daily contexts of African tailors and businessmen, but not in a formal learning setting. The general point seems to be that (mathematical) learning or thinking is contextual in any living culture; it lives and develops and it is used in particular cultural contexts. The often decontextualized (use of) knowledge in the Western school setting is then felt as alienating and foreign by subjects. In my personal field experience with Navajo Indians, the silent refusal by children and native teachers alike of the school and its view on knowledge was striking. The experience with Turkish immigrants gave me a different impression; for them the school is felt to be an alienating but necessary instrument to reach a better position in life. In their case, the heavy emphasis on rote learning (which at the least correlates with the Koran school paradigm of learning) appears to me as a compensation for the lack of insightful learning. At the same time, the present understanding in my country about the lag in school results between native (Flemish) and immigrant (Turkish) children is explained also by the different familiarity in each group with the institution of the school. For example, we find in our research that the immigrant children continue to have duties at home (e.g., washing dishes, caring for newborns, etc.) after they finish school, forcing them to postpone homework until late in the evening; their parents are either unable (because they are illiterate in the Flemish language) or unwilling (because 'that is the task of the school') to help them with tasks from school; the parents motivate their children to perform well in language classes but much less or not at all in mathematics classes, and so on. These and similar findings seem to suggest strongly that the school as an institute must be felt as doubly alien by immigrant children; it has little or nothing to do with life in their 'home culture' and it is a foreign world they have to confront basically all by themselves. Motivation turns out to be problematic and special programmes (geared to overcome these problems of retardation in school) are hence often appreciated as indicators and continuators of deficiency rather than as help for the learning difficulties of the children. Such, at least, are the provisional results of the researches we have been conducting over the last couple of years.

Thus, although the appreciation of the school as a setting for learning is different in both groups (doubt or rejection for the Navajo, and minimal adaptation and frustration for the Turkish immigrant), in neither case is the school really integrated in the culture.

The general study of all those aspects of an institutional nature of mathematics learning is being focused on most of all in a recent and still very modest subdiscipline called 'ethnomathematics' (D'Ambrosio 1985 and 1987). One aspect of the problem area of the institutional aspects of learning can be identified as the uniformisation-through-schooling. The impact of the school as an institution for learning in general (and for mathematics learning in particular) has been assessed by a host of studies. A telling example is the research by Osgood and his collaborators (1975). They found that college students from widely different countries actually use the same basic dimensions or parameters in judging (in their experiments judging the connotative or 'emotional' meaning of words). Another well-known example on the uniformisation influence of schooling is a side result of Scribner & Cole's (1981) research on the effects of adult literacy programs in Africa. They found that on disappearance of other aspects of literacy (such as writing skill, cognitive mastering of the environment, and so on) the structuring of interpersonal relationships in terms of authority which was learned in the school system, remains with the subjects. Mathematics education is too often and too easily identified as part and parcel of the school system of learning and teaching. In that case, it threatens to be alienating rather than emancipating in some cases. This is the worry that is voiced by Third World scholars like D'Ambrosio (e.g., 1985). Hence, the revolutionary thoughts such authors advance to open up the curricula and the teaching programmes to integrate indigenous mathematical notions and activities, 'street mathematics' and the like. In the West and in Third World countries this programme comprises also the introduction of the pocket calculator and of the video game in the classroom, since children are familiar with these items of technology and can more easily learn the basic mathematical knowledge they need through them than through the more traditional and fully 'schoolish' procedures and means. In a more modest way Lancy (1983) advocates a similar openness in 'small' cultures such as those of Papua New Guinea. The minute detail of this study can not be rendered adequately in this short chapter, but Lancy's identification of the problem areas can briefly be pointed out. He distinguishes between three stages of development in the child's growth to adult thinking. The first stage is basically maturational and concerns the primary development in a preschool setting. On the two higher (and later) stages of development, and notably where mathematical knowledge is gradually formed and elaborated, the school format of learning is replacing the traditional ways and perspectives, also in the small cultures. This leads Lancy to point to a growing problem in these cultures, which is inexorably linked with schoolish transfer of knowledge:

> While the Stage II and III processes associated with schooling may be necessary for the production of airplane pilots, they are unnecessary or even harmful for the production of subsistence horticulturalists. How far would a child progress in mastering the village way of life if he firmly believed that answers are found in books, that problem-solving is an individual, intellectual activity, that effort is always and promptly rewarded, ... and so on? (Lancy 1983 p. 210-211).

This pressing question may make us aware of the multifaceted impact of the institution of the school on the cultures of the pupils and on the world in general. It may even be read as an implicit advice towards (partial) deschooling of society. Of course, both moral and intellectual choices are involved here, neither of which I can treat in the present contribution.

CONCLUSION

In this chapter I advocated that the 'empiricist' view on mathematical knowledge is a sound one, at least from the perspective of the anthropologist. I highlighted four parameters in this problem area which, on my view at least, are understudied so far. Each of them has decidedly culture-specific features: the cultural styles of learning; the structure of the vernacular of a community; the contents of knowledge;and the institutional aspects of knowledge transfer (notably in the school setting).

For each of these my suggestion is to adopt culture-specific emphases in mathematics teaching, leading to a situation-bound and necessarily pluralistic perspective on mathematics classrooms.

REFERENCES

D'Ambrosio, U.: 1985, *Socio-cultural Bases for Mathematics Education*, Brasil:Unicamp.

D'Ambrosio, U.: 1987, *Etnomatematica*, Brasil:Campinas.

Bishop, A.J.: 1988a, *Mathematical Enculturation* , Dordrecht:Kluwer.

Bishop, A.J.: 1988b, 'The Interactions of Mathematics Education with Culture', *Cultural Dynamics* , I(2), 145-157.
Carroll, J.B. (Ed.): 1956, *Language, Thought and Reality. Selected writings of Benjamin Lee Whorf* Cambridge Mass. :M.I.T. Press.
Closs, M.P. (Ed.): 1986, *Native American Mathematics* , Austin:University of Texas Press.
Cole, M., Gay, J., Glick, J. and Sharp, D.: 1971, *The Cultural Context of Learning. An exploration in experimental anthropology* London:Methuen.
Cole, M. and Cole, S.: 1989, *Children and Development* , New York:Freeman.
Crump, T.: 1990, *The Anthropology of Numbers*, Cambridge:Cambridge University Press.
Davis, P.J.: 1989, Applied Mathematics as Social Contract In *Mathematics, Education and Society* , Unesco STE Document 35, 24-27.
Davis, P. J. & Hersh, R.: 1981, *The Mathematical Experience* , Boston:Birkhäuser.
Davis, P. J. & Hersh, R.: 1986, *Descartes' dream: The world according to mathematics* San Diego:Harcourt Brace Jovanovich
Dawe, L.: 1988, 'The Impact of Culture in the Mathematics Classrooms of Multicultural Australia', *Cultural Dynamics,* I(2), 196-209.
Ernest, P.: 1991, *The Philosophy of Mathematics Education,* London::Falmer Press.
Frankenstein, M.: 1989, *Relearning Mathematics. A different third R-Radical Maths,* London:Free Association Books.
Ifrah, G.: 1985, *From One to Zero. A universal history of numbers* Harmonsworth:Penguin.
Kearney, M.: 1983, *World View,* Novato Cal.:Chandler & Sharp Publishers Inc.
Keitel, C., Damerow, P. and Bishop, A. J. (Eds.): 1989, *Mathematics, Education and Society,* STE Unesco Document 35, Unesco, Paris.
Lancy, D.: 1983, *Cross-cultural Studies in Cognition and Mathematics,* New York:Academic Press.
Lave, J.: 1988, *Cognition in Practice: Mind, Mathematics, and Culture in Everyday Life* Cambridge:Cambridge University Press.
Lewis, D.: 1972, *We, the Navigators,* Canberra:Australian National University Press.
Needham, J.: 1965, *Science and Civilization in China, Vol.II,* Cambridge:Cambridge University Press.
Osgood, C.E., May, W.H. and Miron, M.S.: 1975, *Cross-cultural universals of objective meaning,* Urbana, Ill.:University of Illinois Press.
Pinxten, R. and Farrer, C.: 1990, 'A Comparative View on Learning', *Cultural Dynamics,* III(3): 215-228.
Pinxten, R., van Dooren, I. and Harvey, F.: 1983, *Anthropology of Space. Explorations into the Natural Philosophy and Semantics of the Navajo,* Philadelphia:Pennsylvania University Press.
Pinxten, R., Soberon, E. and van Dooren, I.: 1987, *Towards a Navajo Geometry,* Gent:KKI Publishers.
Pinxten, R., mimeo ' Wiskunde voor migranten? Een voorstel over wiskunde-onderwijs en kultuur', paper for the Psycho-Medical Centers of Belgium, 1991.
Scribner, S. and Cole, M.: 1981, *The Psychology of Literacy,* Cambridge, Mass.:Harvard University Press.
Tyler, S.: 1978, *The Said and the Unsaid: Mind, Meaning and Culture,* New York:Academic Press.

Witherspoon, G.: 1977 *Language and Art in the Navajo Universe,* Ann Arbor:University of Michigan Press.
Zaslavsky, C.: 1973, *Africa Counts,* Boston:Prindle, Weber & Schmidt.

Rik Pinxten,
State University of Ghent,
Department of Anthropology,
Blandijnberg 2,
9000 Ghent BELGIUM

L. GATTUSO

WHAT HAPPENS WHEN ROBOTS HAVE FEELINGS?

Some years ago, while conducting an investigation with mathophobic students (Gattuso & Lacasse 1986, 1987, Gattuso 1987) I was amazed by the perception the students had of their mathematics teachers. The subjects were 15 college students enrolled in "Mathophobia" workshops. Data were taken from autobiographical questionnaires and from recordings of discussions going on between the pupils and animators during the workshops. Analysis showed very clearly that these students took this equation as a tautology :

"Math teacher = rational = robot = non human"

According to the students, mathematics teachers have a very good mathematical knowledge and an ability to "waltz with number" but they appeared "rational, cold, hurried and not very communicative". Drawing on an image of the human being as having two connected sides such that emphasising one weakens the other *(called in French "le principe des vases communiquants")*, one student said "they had to work hard to be good at maths so they could not develop their human side that much". It was also clear that they strongly associated a mathematics teacher with mathematics itself.

The analysis of these remarks together with observations gathered during the workshops led to the formulation of some hypotheses. If mathematics teachers were to communicate their mathematical experience more openly and be transparent about their activities, their inquiries and their reflections while solving problems, then students might see that the teachers' processes are similar to theirs; they search, hesitate, try out, verify, etc. Occasionally adding historical anecdotes might also help to make mathematics more human. These notions were part of 13 working hypotheses developed as a result of this investigation. The hypotheses were grouped around four dimensions:

1. Affective aspects vs Ability to communicate

(1) The teacher has to put in place communication channels, (2) address the affective dimension in the learning of mathematics and (3) make sure that the students are able to express themselves about their own experience of mathematics

S. Lerman (ed.), Cultural Perspectives on the Mathematics Classroom, 99–114.

2. Relations between students vs mathematics learning,
(4) The teacher should privilege student to student exchanges, (5) free exploration, in groups, seems to be an important factor of learning and it should be possible for the students to experience it and (6) there should also be the possibility for the students to verbalise their mathematical processes.
3. Relations with the teacher vs mathematics learning:
(7) The teacher has to transmit her or his own experience of mathematics, (8) the teacher must have opportunities of supervising individual learning and (9) she or he should multiply the moments of awareness.
4. The relevance of mathematics:
(10) Historical and cultural contributions of mathematics should be adressed in the classroom, (11) the student has to be able to link certain problem-solving procedures to her or his own daily experience, (12) the value of mathematics must be transmitted without mystification and (13) to encourage the student's interest, the mathematical environment has to be concrete, real, and human.

The working hypotheses aimed to create conditions for a desirable level of mathematical learning as well as minimizing the occurrence of negative reactions to mathematics. Clearly, it was necessary to verify their validity in the classroom, and for that a pedagogical approach using these hypotheses had to be defined and tested in the classroom.

Still, it would be interesting to know why is it that mathematics teacher reflect this image? Does it come from their behaviour? Is it that mathematics teachers are trying seriously to be "objective" thus appear "cold" and without any "emotions"? Or do they think it is wrong to have feelings? Maybe mathematics is so constraining that they restrain the expression of feelings? Because of the strong association between the student's perception of mathematics and of their mathematics teacher, this issue is important and would have to be looked out.

AFFECT AND MATHEMATICS

The importance of the affective domain in the learning of mathematics has been noted by many authors. In a study conducted by Nimier (1976, 1977), for example, sixty students' interviews were examined from a psychoanalytical point of view and a resulting questionnaire was administrated to 700 students. The results showed that mathematics was seen as an object of "anguish" or as a defence from anguish. Mathematics cuts people off from the world, disciplines the mind, brings power or creativity depending on the individual personality . A number of other studies on the influence of affective variables in the learning of mathematics have been reviewed by Aiken (1970) and Reyes (1984). More recently, we find various studies dealing with affective factors and their relationship to the cognitive processes involved in problem solving (Mcleod & Adams, 1989).

WHAT ABOUT THE TEACHERS?

Teachers, as problem-solvers, experience affective reactions and have to deal with affective issues in the classroom. In fact, Adams (1989) reports that students' affective responses influence her instructional decisions. Evidence that there are important links between the way teachers organize and conduct problem-solving lessons, and the affective responses and tendencies of their students, is revealed in a study focusing on teaching practices and student affect in problem-solving lessons (Grouws and Cramer 1989). The relationships are complex; however "certain teaching conditions and structures may influence students affective responses to problem-solving" (Grouws and Cramer 1989 p. 160).

Nimier's study (1988), repeated with teachers (30 interviews, 908 questionnaires), revealed that the mode of relations to the mathematical object is firmly fixed in the teacher's personality and, hence, connected with their relationship to their students in the teaching situation.

These processes are mostly unconscious. Although the teacher in her or his instructional decisions is constrained by external factors relating to the class or the institution, she or he is also constrained by internal forces residing in the unconscious. These internal constraints are, according to Blanchard-Laville (1991), important in the teacher's professional decisions. Such internal tensions are not directly accessible except by the use of a psychoanalytical approach conducted by the teacher in some favorable environment as suggested by Blanchard-Laville.

As can be seen, the existence of affect in relation to mathematics is unquestionable and as mathematics teachers we have to take account of it. There is, however, a part of affect which is less permanent and consequently is more difficult to detect; that is the emotional aspect. However, emotions and feelings are present and do affect instructional behaviour.

It is this aspect of affect which I experienced. Following the lines determined by our former research, I developed an investigative pedagogical approach where the main objective was to create an environment in which, through genuine mathematical activities, the affective aspects of learning mathematics would be recognized and coped with (Gattuso & Lacasse 1989). As the experimenter I posed the question: how many of my *a priori* conceptions of teaching do I really put into practice?

One event in particular illustrated the problem. A group of students were 'fooling around' rather than working, and disrupting my concentration. At the first opportunity I went to the board to answer a question. Talking very calmly and explaining the steps, I remarked that "evidently,... as everyone can see..." as if everything was absolutely obvious when it clearly was not. It suddenly occurred to me that my feelings had led me to take control of the situation and that if I were to go on in this way, I could easily crush the students' self-confidence, achieving completely the opposite of the aims of the course.

After the class, I wondered how many of my views concerning the teaching of mathematics were really put into my practice. I realized that my conceptions did not transfer automatically into my daily practice but the fact that an emotional reaction was the cause did not strike me before I got at the point of the analysis in the following study. Was this an exceptional event and which were the causes of discrepancies between conceptions of teaching and the enacted teaching itself were the questions that triggered me? They were important questions because I was trying out a new teaching approach which would in fact be void if I always reverted to my traditional way of teaching.

CONCEPTIONS AND PRACTICE

It is known that although teachers' actions are substantially determined by their thought processes, social constraints and opportunities affect them (Clark & Peterson 1986, Ernest 1991). Case studies conducted to investigate the conceptions of mathematics and mathematics teaching held by three junior high-school teachers showed that the teachers' conceptions of mathematics and mathematics teaching played a significant role in shaping their instructional practice (Thompson 1982, 1984). But beliefs do not necessarily transfer into actions. A study focusing on a mathematics teacher's beliefs about problem-solving revealed conflicts between his expressed beliefs and his classroom practice (Cooney, 1983, 1985). These differences may be due to a different interpretation of the term problem-solving by the teacher and the investigator or by the pressure of the milieu of classroom life. Physical setting or external influences such as expectations from the principal, the community, the parents and the students can have an important impact on teacher behaviour. The teachers' practice is also biased by the curriculum or textbook adopted, as well as by institutional assessment.

In my own study I was aware that some external constraints such as curriculum, lack of an adequate textbook or teaching materials, external assessment or hidden pressures interfere between conceptions and practice causing some of the discrepancies observed. My interest was to examine the discrepancies more closely and learn more about where and why there are discrepancies between conceptions and practice, especially to look out for causes residing in the classroom itself. I had seen that feelings could interfere but my assumptions were that apart from environmental constraints, the students reactions to a new pedagogical approach would be an important cause of interference, or again, that the old and usual type of teaching behavior would overcome expressed conceptions. Therefore, I decided that it would be important to attempt to minimise external causes as much as possible.

THE STUDY

Some of the disparities between conceptions and practice could be due to the use of research methods themselves, for example with a teacher being observed by the researcher. The researcher may interpret the teacher's conceptions in a different way or the teacher's behaviour might be influenced by observation. I assumed that being the teacher and the researcher might resolve these questions. To minimise these factors, and inspired by Schön's "Reflective Practitioner" (1983), I decide to examine my own teaching practice. Since conceptions about teaching would be mine there was no risk of misinterpretation and exterior observation would not interfere. Even though it was clear at first that the best way to get directly to the conceptions was to be the subject and the researcher at the same time, finding a convenient methodology to accomplish the study in these conditions was not easy. First (this has to be a personal choice), being the subject places the researcher in a position to see a lot but also it puts the subject in a very vulnerable position and requires some kind of security. Indeed, to assure some external reliability it is recommended that the researcher be transparent about her work. In fact, in qualitative research we should talk about transparent subjectivity instead of objectivity or neutrality so it is necessary that an other researcher has sufficient documentation to be able to judge of the conclusions of the study in regard to the facts and observations collected (Goetz, LeCompte, 1984, Merriam, 1988). Being reluctant about this openness might endanger the research. This choice proved to be also very rewarding. Being the experimenter and the observed subject gave me access to much more information than is usually available to an external observer who would never have the time and the opportunity to get to know the environment, students, classroom material as deeply.

With both subject and researcher being the same person, however, the question of objectivity and overall validity arises. To deal with these problems I used 'solid' data that could be interpreted by a second researcher. These included tape recordings of the classes and my teaching journal.

Another methodological problem encountered was how would we establish the conceptions about mathematics teaching without influencing them? Others researchers facing the same problems have raised the question of the effect of the interviewer or the questionnaire (Robert & Robinet 1989a, 1989b). They chose to study textbooks and teachers' journals. I choose to do it by analyzing my own work on the subject of mathematics education. All these publications were also available to an external researcher. In this particular case, establishing the conceptions was a relatively easy task and the result can easily be verified.

The core of the study was the confrontation of the practice with the conceptions previously found. It was found necessary to have another researcher to corroborate the findings and in this way to validate the analysis. Hence, the data were recorded in the form of a journal and in the form of some audio recordings, thus permitting the comparison of the practice with the conceptions by an external researcher. It was important to see if my interpretation would be biased to my advantage or against it. The juxtaposition of both analyses revealed no profound contradiction; however it clearly showed that my analysis was reaching more deeply.

The context of the research was also favourable to a reduction of constraints raised by previous studies. Most of them have been conducted with subjects lacking experience such as pre-service or beginner teachers, or those without specific mathematical training (Bromme & Brophy 1986). The lack of experience may in part explain the inconsistencies between conceptions and practice; the beginner may have less precise conceptions or may be without the means to put them into practice or to resist the surrounding social pressures. Without enough mathematical knowledge teachers rely upon textbooks, and the constraints of school syllabuses and examinations have a further important effect (Bromme & Brophy 1986, Dörfler & McLone 1986). At the time of the study, I had been teaching mathematics in college for about twenty years at a level equivalent to British Further Education colleges and Access courses. The context was very familiar to me, thus lessening the pressures due to the school environment. My university degree was in mathematics and I was well acquainted with the content of the course - a remedial course dealing mostly with basic mathematics, analytical geometry and trigonometry.

Pressures due to curriculum, textbooks or other materials were almost nonexistent because I had designed this pedagogical approach and created all the didactical material available to the students. Finally, in Quebec the assessment is the individual teacher's responsibility.

THE TEACHING CONTEXT

The course was a remedial one offered to students who entered college without the necessary background in mathematics. Some of them have difficulty with the subject-matter itself, their preparation being inadequate;

others may lack the necessary skills in terms of working habits; yet others may have a very negative feeling towards mathematics that might stem from past experiences with failure or simply from a deep misunderstanding about mathematical activity. Previous experience pointed out that there were many problems in teaching such non-heterogeneous groups. While mathophobia is present, there is also an expression of indifference towards mathematical learning and, for that matter, towards any kind of learning in general. Also, lack of knowledge of some of the basic skills or concepts in mathematics has a direct impact on subsequent performance of students. Rich and stimulating activities that would permit the students to experience success were needed to overcome their deep-rooted indifference towards mathematics.

An investigative problem-solving approach close to the style adopted in the original workshops was designed: students worked in small groups on problem-solving tasks; very few lectures as such were given, and then only as follow-up or summary; help from the teacher was given as needed; verbal exchanges among the students addressing content, strategies, attitudes, and feelings were encouraged. The content was chosen to allow exploratory activities based as much as possible on hands-on material. This helped them to make sense out of the mathematical activities and gave them the opportunity to verify their solutions.

DATA AND ANALYSIS

Focusing on the discrepancies between conceptions and practice it was important to firstly find a way of conceptualizing my views on mathematics teaching and secondly a means to observe the practice.

Personal publications on mathematics education served as source to assess my previous conceptions. My journal and recordings of classes during the whole semester for two different groups supplied data about my practice.

A list of conceptions was extracted from the papers by two researchers (including myself). The conceptions were written in terms of instructional behavior: "asking questions; supervising the rhythm of the students' work; synthesizing the results; providing opportunities for exploration and discovery; encouraging verbalisation of their mathematical processes, making team-work possible; diagnosing difficulties from their work; underlining students' ideas; leaving responsibilities to the students" etc. Some seventy categories were listed, amongst which 16 concerned the didactical material and four were added during analysis. This list served as a base for the analysis of the practice. My main research question was: "How far are these conceptions reflected in practice and how can the differences be explained ?" I went through the recordings and the journal, coding the practice in terms of these categories rating them with a +, or - whether there was a correspondence or a contradiction between a certain conception and an act of practice. I marked a ± when it was neither a correspondence nor a contradiction. For example one of my conceptions was "responding to a question by a question", but when a student asked a question such as "What is a median?", answering by a question seemed inappropriate. A parallel examination was undertaken by a second researcher for corroborative purposes.

THE OUTCOME

The results strongly suggest that if most of the social constraints are eliminated, the teacher's conceptions about mathematic teaching can largely be transferred into practice. Some 81% of all the observations of behaviours conformed with one of the identified conceptions; the remainder were divided almost equally between conflicting or problematic situations. This figure is an average, however, and not all the conceptions were equally reflected in practice. The discrepancies observed between some of the conceptions and instructional practice can be accounted for by various factors; some are external, some depend of the students, others come from the conceptions themselves, and other revolve around the teacher.

External pressures cannot be completely eliminated. Time-tables, for example, have an important effect; the planning of an activity cannot be the same for a class of one, two or three periods, supervision of students' work is more difficult at the end of the school day. Comparison with other classes cannot be totally avoided.

Other difficulties came from the students. Despite the fact that the class organization favoured team-work, students did not interact as much as expected and the task of stimulating discussion was demanding. Supporting the expression of their mathematical processes using correct mathematical language was compounded by the problem the students had expressing themselves generally. Although the class provided space for individual working pace, some students were slower than most and required me to give more explanation than 'guiding exploration' would require.

The complexity of the subject matter or an ambiguity in the formulation of a mathematical task also led me, at times, to be more directive than supportive, thus deviating from my overall teaching strategies.

An important result was finding that at a certain point some conceptions could be in conflict with each other, thus rendering their use more problematic. For example, three of the conceptions were 'individual teacher supervision', 'encouraging the student to assume responsibility', 'encouraging individual autonomy'. However, if 'individual teacher supervision' is to be important it may impede the assumption of responsibility and the development of autonomy by the students. In fact the analysis reveled that some of my teaching practice in relation to these conceptions tended to be closer to nursing!

Two others categories asked the teacher to 'set the frameworks', giving the students precise instructions about how and what they are expected to do, and to be 'flexible and open-minded', being able to accept students ideas, willing to explore whatever come up, not imposing previously worked-out solutions. Although the first refers to the organization of the activity the other is directed towards the mathematical content. Being in the process of experimentation and trying to follow my own views, I unconsciously overlapped the two conceptions. I was flexible about the organization, becoming at times too permissive, letting the students argue about the test schedule, homeworks and other demands; this was inconsistent with the first conception which required me to give precise directions. The 'conflictual overlapping' found between some of my conceptions was not anticipated because theoretically these were not contradictory. It was their enactment that caused problems. It is probable, however, that awareness followed by caution would help to improve consistency on these points.

Finally, being the subject and the researcher gave me an unexpected insight which was supported by my professional journal and the tape recordings. Looking for explanations of the inconsistencies, I was forced to admit that personal feelings have to be taken in account.

FEELINGS; A CONSTRAINT BETWEEN CONCEPTIONS AND PRACTICE

Although I had previously observed discrepancies between my professional behavior and my conceptions about the teaching of mathematics, I had not realised at the time the role of feelings. This finding was almost accidental. While I was listening to the recordings, I noted down all that came into to my mind. Very early in my work, I adopted a provisional code 'personal reactions' and commented on the lines of: "This is not one of my conceptions about the teaching of mathematics, but it is there. . . " Every time it occurred, I marked it down. In the second stage of the analysis, while I was going through all the transcripts in parallel with my notes, it occurred to me that personal feelings were part of teaching and sometimes interfered between conceptions and practice.

Feelings related to teaching vary from anxiousness to pleasure. Some are negative feelings, others are good, but even if they occasionally obstruct the realization of one of my conceptions the result is not necessarily faulty. Here are some examples that I observed:

(1) At some times during the semester, I felt anxious. Before the first meeting with the students, I wondered about how they would react to a new

approach; during the semester, I wondered if the material would be completed; before a test, I asked myself if they were really ready. I worried because the students did not engage in team work as I expected, I questioned myself about how much of the curriculum they had really learnt. But two of my teaching conceptions said: 'encouraging the student to assume responsibility', 'encouraging individual autonomy' and the analysis of the practice revealed some discrepancies in regards to these conceptions. My anxiousness led me to do too much, bringing to class equipment students should have (compass, protractor,. . .), offering to retake a missed test before the student even asks, etc. This left less opportunity for the student to assume her/his own responsibility and develop autonomy.

(2) Feeling preoccupied can also divert you from your aim. During the firsts weeks while I was going from one group to the other, I was so preoccupied with learning the students' names and getting to know them individually, that I neglected my principle: 'supervision of the student's work'. I took time to go around the class guessing the students' names but meanwhile my attention was not focused on their work. To some degree it was rewarding because after a couple of weeks, I could name them and then really give them individual recognition and attention. But whatever are the causes of preoccupations: personal troubles; difficult students; administrative or curriculum pressures, they can be disturbing and affect teaching efficiency.

(3) Dislikes also play a subtle role. I do not like preparing tests. My feeling is that assessment should be thought over carefully and my opinion is that the traditional paper test evaluates more the capacity for replication than true problem-solving. I do not agree with this approach so that even if at the time of the research I was putting a lot of time into the preparation of the class material, I gave less importance to traditional test preparation, resulting in errors that rendered some of the problems trivial or impossible. My unhappiness with traditional assessment led me to give less attention to this task relative to others. One may be unwilling to teach some particular content or hesitant to engage in problem-solving. Acknowledging dislikes is necessary for a teacher to cope with them otherwise it might affect the teacher's didactical choices.

(4) At other times, some students' behaviour unnerved me, certain attitudes depressed me. I felt irritable. If students went on chatting, I reacted coldly or sarcastically, instead of supervising their work and encouraging them and guiding them in the pursuit of their investigation as expressed in my conceptions. Again after I had tried to explain the aims of a certain task to a student who did not show any interest in the mathematical activity but only in the marks he might get for doing the work, I did not feel like taking time to help him make sense of the activity. I just gave a 'mechanical answer': "do

this and do that". These reactions were totally incompatible with my views on teaching mathematics.

(5) Since at the time I was the only teacher working with this approach I often felt isolated and I found myself drawn into discussions with the students on my work and my studies, but also on other subjects. Solitude as a teacher added to anxiety about the validity of this approach led me to 'give way' instead of being clear in my demands. This fact became evident after some teachers joined me to extend the research.

(6) Fatigue has diverse consequences on my practice. At times this leads to reactions one would expect: I feel impatient, I'm ironical or I suddenly add demands. However, what is clear is that when I feel tired, I tend to teach in what could be called an 'energy saving' style. I show less zeal and I stimulate the students' activities less. I let go, the rhythm of the class is slower, I neglect the 'supervision of the working rhythm'. Instead of guiding the student's effort, asking questions, underlining an idea or giving a hint to help them pursue their investigation, I react more conventionally; I answer and explain more and more often. On occasions, a question leads me to the blackboard, a position I will occupy for most of the remaining course time. I go back to my former method of teaching, taking control of the situation. This manner spares efforts for the teacher but also gives less attention to the students' individual requirements. Feeling tired makes me less efficient when it's time to find an error or to diagnose a problem or the cause of a misunderstanding. It easier to answer or to give a solution, possibly passing by the real difficulty.

(7) Obviously, unpleasant feelings tend to be more detectable and this description might give a misleading impression. Most of the time, things go well, and pleasant feelings can work in both directions from the teacher to the student, but also the converse. When I feel happy, I laugh more easily and kid the students about their errors. I guide them to discover solutions patiently or I tease them with a relevant question encouraging them to go on. The class atmosphere is lighter.

(8) When students call on me, particularly outside class periods, with specific questions showing me their efforts, I enjoy it, and I happily see to their problems, guiding them so they work them out by themselves. More generally, as we observed in a previous study "enthusiasm expressed by the teacher reflects on the students" (Gattuso & Lacasse 1989 p. 66).

(9) From time to time, I feel encouraged by their results or their work, and I then tend to give them more time to explore their ideas; I question them, hinting at a possible alternative; I underline some of their results helping them to extract a general rule; I give some historical anecdotes. Relaxed, I am more attentive to the students' particular problems. On the whole, I will be more consistent with my beliefs.

(10) Going along unexplored paths always brings its measure of anxiety but it is also stimulating and rewarding. This reflective process did show me some of my weaknesses, but also some positive results, it encourages me to go on with my efforts to improve my teaching.

DISCUSSION

At first, feelings seem to override conceptions, at least the consciously expressed conceptions, but maybe they only play a revealing role, catalysing the expression of more deeply ingrown beliefs. Nimier's studies (1985) exposed the fact that mathematics serves as a support to a variety of unconscious constructs. The way in which a mathematics teacher relates to students is strongly associated with her/his image of mathematics and to her/his personal unconscious needs. For example, mathematics seen as 'laws and rules' can serve as a constraint providing the enactment of desires and emotions or contrarily giving way to the manifestation of a degree of aggressiveness and of intolerance to the students' contradictions, hesitations and doubts; "there is one right way of doing maths..."

Mathematics seen as a structured body of knowledge serves as a means to security but it may also satisfy a domination purpose. Mathematics may also be protective when it is perceived as a peaceful world closed to perturbing emotions. The teacher living this situation will probably prevent all forms of communication with the students. Thus mathematics fulfilling personal unconscious needs can also lead to internal equilibrium that will transfer into the teaching relationship.

Different positions may occasion different answers depending also on the student's own situation. In my case, not taking an authoritative position as a mathematics teacher possibly led students to feel insecure, and a few did. Although previous experience had shown this possibility I did not change my position but I explained frequently to the students why I adopted this pedagogical approach; showing that I knew where we were going filled most of their needs in the matter of security. What mostly appeared from the students' comments was that at last they understood what they were doing, they experienced some pleasure doing mathematics and they did not have to follow blindly previously imposed rules. Since these students were in their late teens, it is understandable that their more adult side rejected authority, even if some childish part of them could still feel insecure.

Feelings intervene possibly causing some discrepancies between the espoused model of teaching mathematics or revealing some deep beliefs about mathematics; feelings are there, they have an impact in the classroom and we cannot ignore them. The fist step is to acknowledge them and in some way to share them. Feeling happy and relaxed reflects in the classroom atmosphere, so negative feelings must also do so. There is probably a lot of energy spent hiding emotions and feelings. Like children (or everyone) students sense the teacher feels bad. For instance, once as I entered the classroom, a student asked:"Why are you mad at us?". The real reason of my humour was a parking ticket! Seeing how wrongly my attitude was interpreted, I volunteered the information and we all laughed about it. Hiding feelings might give wrong impressions. However, we have to be cautious not to fall into the opposite extreme; the classroom is not the place for therapy.

Feelings are there and it is necessary to acknowledge them and to take account of them as teachers, but also as teacher educators and as researchers. Recognizing the existence of feelings is indispensable for the teacher. The teacher's task is often a solitary one, a teacher can experiment with a wide range of feelings throughout their professional life. It is important to know they are a normal part of teaching practice. Once such feelings are recognized the channeling of some of their unwanted effects becomes possible, permitting the teacher to have better management of his/her professional practice. And maybe ultimately, the teacher will feel free to talk about feelings even to the students.

Teacher educators have the responsibility for preparing future teachers to cope with their feelings, knowing that they affect instructional practice. Informed, the new teacher might be more prepared to face the inevitable difficulties of their profession. Besides, let us not forget that teacher educators are teachers themselves.

As researchers we have to take account of feelings too. First, it is important, as I experienced, to look out for reactions and effects of feelings in research, particularly in classroom settings. Secondly, there is a clear need for further studies on feelings and on their effects on teaching practice, and research on ways of helping the teacher to recognize their feelings and to deal with them.

CONCLUSION

Teaching is a very complex task. At the beginning my interest was to find ways to reconcile the students with mathematics. Although the link between affective factors and the learning of mathematics had been established, little had been done in classrooms to cope with them. Even though external factors may also intervene with the learning of mathematics, our previous study (Gattuso & Lacasse 1986, 1987b) persuaded us that a lot could be done in the classroom, and the teacher had an important role. At first, the importance of communication between the teacher and the learner and amongst the learners themselves were the key concepts. However, experimenting with these views moved my focus onto the mathematical task itself, leading me to a problem-solving approach that gave an active role to the student, consequently repositioning the teacher.

The student gains confidence while engaging in a mathematical task as long as it leads to some success. There is actually a great need for such tasks, especially at a more advanced level, and a need to study mathematical concepts from a didactical point of view in order to prepare classroom material and to prepare prospective teachers.

For the teacher to project a living image of mathematics, that is of mathematics as growing and dynamic knowledge, the teacher must reveal his/her own thinking processes particularly in the problem-solving. One should let students explore and be ready to react to various questions. Thus self confidence in mathematics and in teaching is an important aim for the teacher. One of the best preparations would be a solid mathematical background but from the learning and teaching point of view. A lack of mathematical knowledge can lead to a very narrow presentation of mathematics, avoiding 'shaded' areas, but a strong mathematical background without any knowledge about how one learns and how concepts can be acquired can give very poor teaching results.

Overall a great part of teaching mathematics concerns the relationship between persons through mathematical knowledge. That is why it is understandable that personal feelings intervene.

To look into these and others factors that affect teaching, there is a necessity to reflect on professional actions. Teachers need the tools and the support to improve while they are teaching. Reflection must become a familiar process.

Finally let us ask ourselves why this perception of the mathematics teacher as a mechanical non-human robot is so widely accepted.

"Mathematics teachers, robots? Why?"

REFERENCES

Adams, V. M.: 1989, Affective Issues in Teaching Problem Solving: A Teacher's Perspective In D. B. McLeod and V. M. Adams (Eds.) *Affect and mathematical problem solving: a new perspective.* New York: Springer-Verlag. p. 192-201.

Aiken, L.: 1970, Attitudes toward Mathematics *Review of Educational Researc* Vol. 40, No. 4, p. 551-596.

Blanchard-Laville, C.: 1991, La dimension du travail Psychique dans la formation continue des enseignant(e)s de mathématiques In F. Furinghetti (Ed.) *Proceedings of the Fifteenth International Conference of the Group for the Psychology of Mathematics Education* Assisi, Italy, p. 152-159.

Bromme, R. and Brophy, J.: 1986, Teachers' cognitive activities In B. Christiansen, A.G. Howson, and M. Otte (Eds.), *Perspectives on Mathematics Education* Dordrecht: Reidel, p. 99-139.

Clark, C. and Peterson, P.: 198, Teachers' thought processess In M. Wittrock (Ed.) *Handbook of research on teaching, third edition* New York: Macmillan p. 255-296.

Cooney, T.: 1983, Espoused beliefs and beliefs in practice: The cases of Fred and Janice In J. C. Bergeron & N. Herscovics (Eds), *Actes de la rencontre annuelle PME-NA, 2,* Montréal, p. 162-169.

Cooney, T.: 1985, A beginning teacher's view of problem solving *Journal for Research in Mathematics Education* Vol 16, No. 5, p. 324-336.

Dörfler, W. & McLone, R. R.: 1986, Mathematics as a school subject In B. Christiansen, A. G. Howson & M. Otte (Eds.), *Perspectives on Mathematics Education.* Dordrecht:Reidel, p. 49-97.

Ernest, P.: 1991, *The Philosophy of Mathematics Education* London:Falmer Press.

Gattuso, L. and Lacasse, R.: 1986, *Les mathophobes une expérience de réinsertion au niveau collégial* Cégep du Vieux Montréal.

Gattuso, L. and Lacasse, R.: 1987a, Le vécu des mathophobes *Bulletin amq*, Vol. 27 No. 2 mai. p. 33-35.

Gattuso, L. and Lacasse, R.: 1987b, Les mathophobes une expérience de réinsertion au niveau collégial In *Actes du onzième congrès international de Psychology of Mathematics Education, PME-XI*, édité par Jacques C. Bergeron, Nicolas Herscovics, Carolyn Kieran. Juillet: Montréal.

Gattuso, L. and Lacasse, R.: 1988, Intervention in mathematics course at the college level In Andrea Borbas (Ed.)*Proceedings of the Twelfh International Conference of the Group for the Psychology of Mathematics Education* Veszprém. 425-432.

Gattuso, L. and Lacasse, R.: 1989, *Les maths, le coeur et la raison. Un modèle d'intervention dans une classe de mathématiques au collégial* Cégep du Vieux Montréal.

Gattuso, L. & Lacasse, R.: 1990, Les maths, le coeur et la raison *Bulletin amq*, octobre. p. 2537

Gattuso, L.: 1987, Les mathophobes une expérience de réinsertion au niveau collégial Mémoire de maîtrise. *Sciences de l'Éducation* Université de Montréal.

Gattuso, L., Lacasse, R., Lemire, V. and Van Der Maren, J. M.: 1989, Quelques aspects sociaux-affectifs de l'enseignement des mathématiques, le vécu des mathophobes *Revue des Sciences de l'Éducation* Vol. 15, No. 2, p. 193-218.

Goetz, J. P. and Lecompte, M. D.: 1984, *Ethnography and qualitative design in educational research.* Orlando, Fla.:Academic Press.

Grouws, D. A. and Cramer, K.: 1989, Teaching Practices and Student Affect in Problem-Solving Lessons of Select Junior-High Mathematics Teachers In D. B. McLeod, and V. M. Adams (Eds.), *Affect and mathematical problem solving: a new perspective* New York: Springer-Verlag, p. 149-161.

McLeod, D. B. and Adams, V. M. (Eds.): 1989, *Affect and mathematical problem solving: a new perspective* New York: Springer-Verlag.

Merriam, S. B.: 1988, *Case study research in education. A qualitative approach.* San Francisco:Jossey-Bass.

Nimier, J.: 1976, *Mathématiques et affectivité* Stock.

Nimier, J.: 1977, Mathématiques et affectivité *Educational Studies in Mathematics* Vol. 8, p. 241-250.

Nimier, J.: 1985, *Les maths, le Français, les langues...à quoi ça sert?* Paris:Cedic/Nathan.

Nimier, J.: 1988, *Les modes de relations aux mathématiques* Paris:Méridiens Kincksieck.

Reyes, L. H.: 1984, Affective Variables and Mathematics Education *Elementary School Journal,* Vol. 84 No. 5, p. 558-581.

Robert, A. and Robinet, J.: 1989a, *Représentations des enseignants de mathématiques sur les mathématiques et leur enseignement* Cahier Didirem 1, Paris: IREM Paris 7.

Robert, A. and Robinet, J.: 1989b, *Énoncés d'exercices de manuel de seconde et représentations des auteurs de manuels* Cahier Didirem 4, Paris: IREM Paris 7.

Schön, D.: 1983, *The reflective practioner* London: Temple Smith.

Thompson, A. G.: 1982, *Teachers' conceptions of mathematics and mathematics teaching: three case studies* unpublished PhD dissertation, Athens, Geogia: UMI 8228729 02950.

Thompson, A. G.:1984, The relationship of teachers' conceptions of mathematics and mathematics teaching to instructional practice *Educational Studies in Mathematics*, Vol. 15, p. 105-127.

Linda Gattuso,
Département de mathematiques et
d'informatique,
Université du Québec a Montréal,
Case Postale 8888, Suc. "A",
Montréal H3C 3P8 CANADA

C. MORGAN

THE COMPUTER AS A CATALYST IN THE MATHEMATICS CLASSROOM?

> [In the year 2000] . . . In the classroom of the mathematics teacher will be a bank of microcomputers, as many as one for each member of the class, together with a full range of programs. This is one subject . . . where a few children will be able to learn entirely through computers. (Hawkridge 1983, p. 196)

> Although teachers are usually present, their interventions are more similar to those of the expert dancers in the samba school than those of a traditional teacher armed with lesson plans and a set curriculum. The LOGO teacher will answer questions, provide help if asked, and sometimes sit down next to a student and say: "Let me show you something." What is shown is not dictated by a set syllabus. Sometimes it is something the student can use for an immediate project. Sometimes it is something that the teacher has recently learned and thinks the student would enjoy. Sometimes the teacher is simply acting spontaneously as people do in all unstructured social situations when they are excited about what they are doing. The LOGO environment is like the samba school also in the fact that the flow of ideas and even of instructions is not a one-way street. The environment is designed to foster richer and deeper interactions than are commonly seen in schools today in connection with anything mathematical. (Papert 1980, p. 179)

Visions of the transformation of education by computers have been appearing for as long as computers have been around. The dominant idea of those not directly involved in education (and some of those who are) is of the computer displacing the teacher as the main source of education for children or at least delegating her to a managerial or socialising role. Mathematics education is frequently seen as particularly suitable for the application of computerised teaching and learning, probably because of prevailing views about the nature of mathematics itself and the nature of mathematics education. Whether you see this as dehumanising or as liberating depends not only on your beliefs about the nature and capabilities of computers but also on your beliefs about teaching and learning and the relationships between teachers and children in the classroom. However, such views of the effect of new technology represent a deterministic, technocentric model of the way that change takes place - a model which positions the new technology as the driving force of change and does not take account of existing or evolving social relationships within the classroom.

On the other side of the coin is the belief that introducing the computer into the mathematics classroom will not make any fundamental difference to either the relationships within the classroom or the types of activities which

S. Lerman (ed.), Cultural Perspectives on the Mathematics Classroom, 115–131.

take place there. When making their first steps towards using a computer in their lessons, many mathematics teachers see it as a source of activities - like worksheets or books but more motivating (Morgan 1991) - which will fit neatly into their existing classroom practice. While these teachers have only limited access to hardware and software and support in using them, this belief is likely to persist and to be valid to some extent. It has been found that teachers are very good at adapting innovations to fit their existing practice (eg Olson 1980). I will argue, however, that teachers using microcomputers within contexts which encourage reflection on their own experiences may find that the nature of the interactions between teacher, pupils and computer raises issues which challenge their beliefs about teaching and learning mathematics and thus bring about changes in these interactions.

TECHNOLOGY AND CULTURE

Different forms of technology can be seen to facilitate different forms of social relations. The development of machinery capable of mass production hastened the end of the feudal system in Europe. More recently, the increasing computerisation of manufacture has been accompanied by the evolution of different roles for workers involved in industry. To take an educational example, consider the effects in schools of the change from using slates to having a plentiful supply of paper for children's use. This technological change must have had impact on areas such as the valuing and sharing of children's work, the ways in which assessment takes place, the possibility of doing extended pieces of work, and so on.

The particular form of the social relations is not, however, wholly determined by the available technology. The lack of a deterministic relationship between available technology (in a broad sense) and the culture of the mathematics classroom can be illustrated by considering the different ways in which the same mathematics text books or schemes are used in different schools or even in different teachers' classrooms within the same school. Two schools both using the SMP 11-16 scheme with year 7 (11-12 years old) provide examples of the wide variety of different types of use of the same 'technology' with teachers and children taking different roles.

In School T, the children work individually in a mixed ability class, choosing for themselves which booklet they wish to use from those at their level. Occasionally two will choose to work together on the same booklet but on the whole there is little discussion about mathematics; most talk between children is social. The teacher 'goes round' the class responding to requests for help.

In School S, also in a mixed ability class, the teacher has grouped the children into threes and fours according to her perception of similar 'ability' and social compatibility. Each group works together using the same booklet which, as in School T, the children in the group are allowed to choose. Each lesson, the teacher plans to focus on a particular group. She sits with that group for most of the lesson, talking with them and listening while they work. If children in other groups need help, they first ask others in their own group, then children in other groups who have already 'done' their booklet before asking the teacher.

If the 'old' technology with which we are familiar may be used in classrooms where the things that teachers and children do and the ways in which they relate to each other are so diverse, can we assume that the introduction of a 'new' technology such as the microcomputer will necessarily lead to classrooms like either of those characterised at the beginning of this chapter, or even to any fundamental change at all? I would suggest that microcomputers do provide the power to transform both the kind of mathematics that is done in the classroom and the roles that teachers and pupils play. What these changes are and whether they actually take place, however, depend on many other factors including the existing classroom and school culture, the expectations that pupils and teachers bring to the classroom, and the support provided for the teacher during the period of introduction. In considering the possible changes which might take place it is useful to look at the different ways in which computers are currently being integrated into the classroom.

The discussion which follows refers primarily to the use of one or two microcomputers within the usual mathematics classroom. While occasional or even regular mathematics lessons in computer laboratories are also to be found in many schools, they tend to be seen by both pupils and teachers as different and separate from other mathematics lessons (Watson 1990). Their effect on relationships within the ordinary mathematics classroom is, therefore, likely to be limited.

COMPUTERS IN MATHEMATICS CLASSROOMS: HOW MAY THEY BE USED?

Introducing one or more computers into a teacher's classrom is not simply a management problem or a question of acquiring technical know-how; a key problem which often goes unrecognised is the integration of the computer into the teacher's and the pupils' repertoires of acceptable classroom activity. 'Using the computer' is not a single type of activity but may take many different forms, not only because different software may be used but because the same software may be used in many different ways.

There has been much debate about the relative merits of using different types of software (see, for example, Ball, 1990 and various articles in *Micromath*, including a feature in the Summer 1991 issue debating the "short program" question); the argument being mainly polarised between the advocates of using "short" programs and those who extol the educational benefits of what has been variously characterised as "content free", "general purpose", or "framework" software. The use of some types of software, in particular "content free" software and the programming language Logo, has been hailed by enthusiasts as a way of changing the type of activities and learning that take place within the classroom. On the other hand, advocates of the use of "short" programs consider a major advantage to be the ease with which they can be used by both teachers and pupils. This debate echoes the different positions on the effects of computers themselves outlined at the beginning of this chapter. Both sides are taking a technocentric deterministic view of the piece of software's effect on the type of educational experience.

There have been a number of attempts to classify types of computer use according to the kind of software being used. For example, Taylor (1980) distinguishes between uses where the computer acts as Tutor (providing information, questioning and responding to answers etc), as Tool or as Tutee (when the child programs the computer herself). Such attempts are not, however, completely successful not only because it is sometimes difficult to decide which category a particular piece of software falls into but also because the type of activity is not entirely determined by the particular piece of software being used. For example, Logo can be used in the "revolutionary" way envisaged by Papert (which falls clearly into the "Tutee" category) or in an "evolutionary" way (Lemerise 1990) which imposes far more structure on children's experiences and appears more compatible with the traditional mathematics curriculum and classroom. Although the child may be in control of the machine when she is engaged in using a programming language, the child's own actions may themselves be controlled by the way in which the task is defined by teacher or worksheet. The ways in which the computer, and indeed any particular piece of software, is used and, in particular, the degree of control the child is allowed, are not independent of the existing classroom culture.

There is an interaction between the software and the roles that the teacher and the pupils play, in particular the location of control within the classroom. I have identified four categories of use which are distinguished by the relationship between teacher, pupil, computer and mathematics. In three of these categories the child's degree of control over her learning activity lies on a continuum from being directed to directing herself. The fourth category of use, the computer as a resource, is closely related to Taylor's "Tool" usage; the focus of the child's activity being on a problem arising away from the computer. The degree of the child's control over her activity during this type of use may, therefore, lie on any part of the continuum depending on the non-computing constraints of the activity. Some pieces of software can be used in ways which fit into more than one category, even simultaneously; others are structured in ways which limit the ways in which they can be used in the classroom.

MODES OF COMPUTER USE

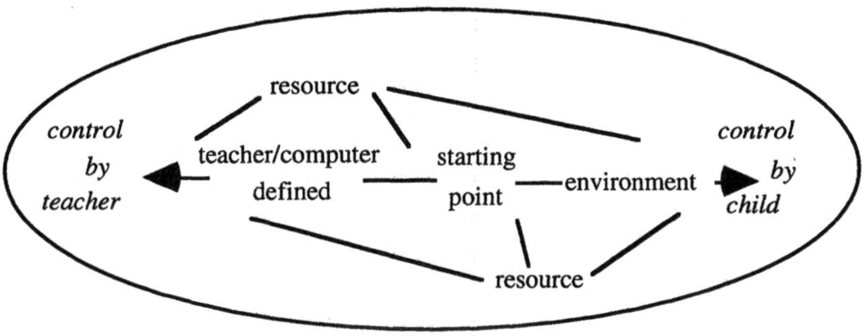

Computer Or Teacher Defined Activities

These are activities in which the teacher (or book, workcard or computer program) maintains control over the direction of the child's learning experiences by posing closed questions and guiding the pupils through a predetermined sequence of experiences (not necessarily the same sequence for every pupil). Much of the earliest mathematics software could only be used with the computer in complete control. "Drill and practice" is easy to program on simple machines and also fits in well with "popular" views of what mathematics education should be like, with the emphasis on "basics". Its use has not been widespread in mathematics classrooms; I suspect this may be because those teachers most likely to take on the innovation of computer use were less likely to feel that extended drill and practice was a respectable classroom activity. (The relationship between teachers' beliefs about mathematics teaching and their willingness to make such innovations in their classrooms would be worth investigating.) Nevertheless, there are still many such programs aimed at the home computer market. The "programmed learning" type of software (eg PLATO) usually associated with behaviourist theories of learning also exemplifies an extreme form of this kind of activity and is deliberately designed to maintain strict control over the type and order of the experiences met by the user. Again, this style of program has not established itself to any great extent in British mathematics classrooms; there are, however, other types of software coming from sources less than sympathetic to behaviourism which nevertheless may be used by teachers in ways which severely restrict the choices available to the pupils.

Many (though not all) of the SMILE programs, for example, are easiest to use in ways which come into this category.

e.g. The program GUESSD is presented as a game for the child to play against the computer. A target number is chosen by the computer. The child makes successive guesses and is told whether they are too big or too small. There is no way in which the child can change the structure of the activity. Even the difficulty of each successive target number is set by the program according to how successful the child has been.

Whether such programs are integrated into the scheme of work with specific mathematical learning objectives or they are used as "games" at the end of a lesson or as a reward for completing other work, once the child is using the program she has extremely little freedom to change the parameters of the activity but must follow predetermined paths. While some software can only be used in this way, even "content free" software such as Logo or spreadsheets can become completely teacher defined when used in conjunction with structured worksheets or other detailed teacher directions.

Computer As Provider Of A Starting Point

Some computer programs can be used to give children a starting idea which they may then develop following their own paths of enquiry, either using the computer as a tool or working away from the computer and possibly returning to it when appropriate. Examples of programs which are designed to be used in this way include, for example, many of those in the MEP Secondary Mathematics with a Micro Pack and in the ATM's SLIMWAM 1 and 2. In spite of the excellent documentation provided with these packs describing lessons using programs of this type, many teachers have found them difficult to use because they have been unsure of when and how to release control of the direction to be taken to the children themselves. The ideas presented in the documentation have not fitted in with their existing classroom practice and this has led to much software lying in cupboards unused. In other cases, teachers, by providing the children with worksheets or by setting restricted targets, have structured the course of the children's enquiry to such an extent that the activity has become teacher defined. On the other hand, even an apparently restrictive program like GUESSD, described above, can be used as a starting point for work away from the computer involving discussion and investigation of alternative strategies for finding the target number.

Computer As A Mathematical Environment

This is the type of computer use envisaged by Papert (1980) when he wrote about children 'learning mathematics in "Mathland"', though in most mathematics classrooms at present it happens only on a very modest scale, if at all. Children use the environment provided by the software to explore mathematical ideas, setting their own goals or working towards general goals set by the teacher and devising their own strategies to achieve those goals. Interaction with the software involves the children in mathematical thinking, although the inevitability of this involvement may be questioned (see for example Hillel et al (1989), Hoyles & Noss (1987) about Logo; Goldstein (1990) about mathematical adventure games).

The programming language Logo is the most commonly given example of software that provides a mathematical environment. Some mathematics teachers now, however, express unwillingness to become involved with using Logo or even hostility towards the idea, claiming either that there is not time for the children to learn programming or that they cannot see what mathematical benefit would come from it. The ambitious claims of Papert and even the less ambitious claims of others who have worked for extended periods of time with children using Logo (eg Hoyles & Sutherland, 1989) seem so far removed from these teachers' classroom experience that they cannot see its relevance for them. This reflects a mismatch between the suitability of Logo to act as an environment in which children control the direction of their own learning and the existing, more teacher controlled, mathematics classroom.

Computer As A Resource

One mathematics department I worked with recently did not have permanent access to any hardware. When they tried to make their case to the headteacher, he told them (only partly in jest) that he would allocate a computer to them only if they could demonstrate that it would be used during every lesson. Maybe not every lesson can involve activities focused around the microcomputer. However, it is possible to argue that it is a resource like calculators, squared paper or counters which has multiple and not always predictable applications in many areas of the mathematics curriculum and therefore needs to be available at all times for children to choose to use as appropriate. This is the position implied in the National Curriculum Non-Statutory Guidance for Mathematics when it recommends that within a scheme of work:

> Activities should provide opportunities for pupils to develop skills in selecting and using a wide range of mathematical tools - constructional kits, drawing instruments, measuring equipment, calculating aids, electronic calculators and micro-computers. They should also help pupils to select with confidence the most appropriate ways to tackle different problems. (NCC 1989)

Unfortunately the introduction of the National Curriculum has not been accompanied by increased funding to enable this to happen.

Teacher H (in a school with a very different policy for computer provision) has considerable experience in using the microcomputer in the way suggested by the NCC, particularly using programming in BASIC to perform repetitive calculations and to plot graphs. The pupils in her classes are used to working investigatively, controlling the direction of their own work and turning to the computer when they considered it to be appropriate. Even when they are unable to do the programming themselves they are often able to identify where it might be useful and will ask their peers or their teacher to provide a program for them. Her confidence with an investigative approach which encouraged both the children and herself to take risks has allowed H to experiment successfully with using the computer in similar ways.

The use of the computer as a resource may occur during teacher directed activities or while children are directing their own work. In order for extensive use of this type to occur, however, either the teacher must have considerable experience and knowledge of what the computer and the software can do and where it might be useful (in order to direct pupils to use it) or the classroom culture must accept that the teacher may not be the only source of expertise and allow children the freedom to experiment.

COMPUTERS AND CHANGES IN THE CLASSROOM

> The most exciting part of the educational computing enterprise will be its effect on classroom culture: on attitudes and atmosphere, on the patterns of intervention, and on the location of control in the classroom. (Weir 1987, p. 246)

The categories of computer use described above can be seen to parallel teachers' ways of working within the mathematics classroom with or without computers. The teacher whose classroom activities are firmly planned and directed is most likely to use the computer in similar, clearly defined and controlled ways while the teacher who prefers an investigational approach and allows the children to make decisions about the direction of their learning is more likely to favour computer activities which mirror this style. The introduction of computers by itself does not necessarily disturb teachers' classroom practice (Wright 1987, Mehan 1989). As I have argued elsewhere (Morgan 1990), the widespread popularity of the SMILE programs in secondary and primary mathematics classrooms in the UK is largely due to the ease with which they can be integrated into a variety of different types of classroom practices without challenging the existing relationships between teacher, pupils and mathematics.

At present, most mathematics teachers have only limited access to computers and have received little support in developing ways of using them in the classroom. The struggle to come to terms with technicalities and management problems may be so great and time consuming that it overshadows the possibility of reflecting on and changing existing practice. There is evidence, however, that, where teachers have relatively easy, long term access to computers and when they are provided with opportunities for observation and reflection, their ideas about how children learn and the relationships within the classroom may be challenged (Olson 1988, Mehan 1989, Morgan 1991). There are features of computer use which, within such a supportive environment, facilitate particular changes in the classroom. Most of the following examples which illustrate these features are drawn from observations and discussions with experienced secondary mathematics teachers participating in an in-service programme intended to introduce the use of microcomputers into the teaching and learning of mathematics. This programme is described more fully elsewhere (Morgan 1991). It included some team teaching and collaborative planning and evaluation of lessons as well as more formal centre-based sessions.

The computer makes children's activity and mathematical understanding more visible

When children are engaged in non-computer mathematics, the easiest way for the teacher to assess their understanding is by observing the product of their activity, whether it is a written or spoken answer, a completed graph, diagram or model. There are often, however, several ways of achieving the same product using varying types and levels of mathematical knowledge and understanding. Of course, the processes used may be revealed by close observation or questioning but this is time consuming and may not be effective if the teacher does not ask the right question or the child is unable to verbalise her ideas in a form that is understandable for the teacher.

The public nature of the computer screen allows the teacher easier observation of the pupil's activity without intruding on it. More fundamentally, the formal nature of the child's interaction with the computer makes the structure of her mathematical processes visible as well as the product of her activity.

Teacher M gave a number of pairs of children, all of whom had been successful completing symmetrical shapes with paper and pencil, a list of Logo instructions which would draw half a simple shape and asked them to complete the drawing to make it symmetrical. Most of the pairs completed the task quickly, using the numbers that had been contained in the original instructions. One pair, however, never noticed that this would help but used trial and error throughout to complete the task. Observing this, M realised that this pair, although successful at completing the task, did not have the same sort of formal understanding of the properties of symmetrical shapes that had been demonstrated by the other children. She proceeded to plan different tasks for this pair to help them to achieve such formal understanding.

Different types of activity reveal different aspects of achievement which may prompt teachers to re-evaluate their assessments of children.

Teacher B planned an activity using short programs in BASIC in which children were to input values and use trial and error to discover the rules of linear functions. As B had expected, most of the children in the class who tried this activity found it challenging and each pair took about 20 minutes to reach a point at which the teacher was satisfied. To her surprise, one pair which B had not previously considered to be among the "best" in the class completed the activity far more quickly and was able to explain and demonstrate a general method for finding the rule of any linear function. While she was not sure how to follow this observation up, she did express concern that the work that these children normally did might not be providing opportunities for them to excel.

Of course, discontinuity in a child's achievement does not occur only between non-computer and computer activities. For example, with the introduction in the UK of assessment by a component of "coursework" as well as by traditional timed written tests at GCSE (a public examination taken by most children at age 16+) many cases have been found of chidren who achieve considerably different results in the different modes of assessment. The opportunity to engage in investigational work and practical problem solving, working independently or in groups over an extended period of time, has enabled some children to demonstrate a level of mathematical achievement not apparent in their performance on timed tests. In that case, in spite of fine words to the contrary, the traditional test results have usually been taken to be the "real" measure of the child's mathematical achievement. The current high status of computer use, however, unlike the relatively low status of "coursework", means that the teacher's view of the overall achievement of children like those in the example above may be affected, not only in the specific topic with which they were engaged at the time.

As well as opening the teacher's eyes to unexpected achievements among her pupils, the characteristic of visibility has the potential to change pupils' attitudes towards cooperative working and towards the private/public nature of their work in the matematics classroom. While the relationship between child and computer is sometimes a private one which enables an individual to experiment and make errors in a non-threatening environment, the display can at other times provide a focus for small groups of children to work collaboratively.

Much of the work in many mathematics classrooms involves children in working and recording their results individually. The product of mathematical activity is usually private even when the children are organised in groups working on the same topic. There are few opportunities to share and even when this is possible, children may be unwilling to do so. Many activities using the computer, however, do not involve written records and, while this causes some worries for teachers, the fact that the product of the activity is visible in a public form encourages children to share it.

When discussing his written work in mathematics with a year 10 boy (age 15), I asked him if he ever looked at other peoples maths work or if they ever looked at his. His immediate response was an unequivocal "no" but he then qualified it by saying that when he was working at the computer he sometimes called to his friends "Hey, look at what I've got here".

Using the computer is clearly seen as qualitatively different from most other work in the mathematics classroom and it therefore allows this boy to behave in a different way towards his work and his peers in this context. This raises the question of whether, as children become accustomed to sharing computer work, their attitudes towards sharing other work will also change, thus affecting the individual, private nature of most school mathematics. Hoyles and Sutherland (1989) note that collaboration on Logo tasks during their longitudinal study also spread to "other" mathematics and increased mathematical talk within the classroom as a whole.

The computer can enable children to learn concepts in different orders and at different rates

The computer environment allows children to experience mathematical concepts in different ways which, it has been argued (Noss 1988), makes formalisation natural in a way that the traditional non-computer classroom environment does not. Such formalisation may therefore be accessible for more children at earlier ages and may be planned for by the teacher, thus causing a change in the order of the programme of study for some or all children. I do not wish in this chapter to address the question of what potential the new technology has to change the subject matter of school mathematics; this has been discussed elsewhere (eg Fey, 1989) and is in any case likely to change as the technology and our knowledge of it develop. It is the relationship of the child and the teacher to the subject matter that currently concerns me.

Some children in A's year 8 class (12-13 years old) had been provided with a worksheet showing some graphs and some equations and had been asked to match each equation to its graph. Using Omnigraph exploring linear graphs and their equations by trial and error, they began to develop clearly expressed ideas about gradient and intercept and their relationship to the equation.

A commented that the children:

" . . . can go way beyond the point you would normally have taken them without the computer . . . you wouldn't have developed some of those points until much further up the school. "

She saw this as having implications for the work that she and the other teachers in the department will plan for these children when they are higher up the school, particularly as not all of those who took part in the activity achieved the same level of understanding.

The task that A originally set these children was relatively teacher directed. Their interaction with the computer, however, led them to explore and discuss issues not explicitly related to the task. Together with their teacher's recognition of the value of this and her willingness to allow the children to take over the direction of the activity, this changed the nature of the use of the computer to that of a mathematical environment. The computer used as a mathematical environment provides opportunities for meeting powerful mathematical ideas during the course of the exploration of that environment but it is not possible to predict with certainty what, whether or by whom. This implies that the teacher can no longer play the role of expert source of all knowledge in the mathematics classroom.

CHANGES IN THE ROLE OF THE TEACHER ARISING FROM COMPUTER USE

Olson (1988) identifies two areas in which the computer may cause teachers to rethink their practice: the content of what they teach and their own role in helping children to learn. The examples given above show how teachers, by gaining insight into individual differences in children's learning, are prompted to reconsider the content of their teaching for these children not only within the context of computer use. During the in-service programme referred to above, three major issues related to the teacher's role and relationships within the classroom became important for the teachers involved: grouping children within the class, assessment and intervention. Those teachers who had the opportunity to be involved in team teaching and collaborative planning and evaluation of lessons were more aware of these issues than those who only participated in the formal course and were isolated in their classrooms. Although these issues arose during computer use, they clearly have wider implications for these teachers' practice.

Grouping Children

The customary classroom management style of most of the teachers involved was either whole class teaching with all the children working on the same materials or "individualised learning" using a commercial individualised scheme. Few therefore were used to planning different work for different groups within the class. In order to make effective use of a single microcomputer in their classroom, however, some form of group work became essential. At first, grouping was done more or less arbitrarily as a means of ensuring "fair turns" for all.

While this concern for "fair turns" may initially reflect the novelty of the computer in the classroom and the idea that it is particularly enjoyable for children to use it, it is also related to a techno-centric attitude towards teaching and learning, viewing mathematics teaching as providing a set of activities which will determine the mathematical activity and the learning of the children. It represents a belief that: "All the children in my class learn in similar ways (though possibly at different rates). They therefore need to be provided with the same experiences (not necessarily at the same time)". When there is only one computer in the classroom this stricture has severe implications for the types of computer experiences that it is possible to offer to the children. In particular, the amount of time available for any individual child to work with the computer is very limited and, furthermore, the activity needs to be self-contained and independent of other activities so that each child's "turn" depends only on the availability of the computer, not on her own readiness or need. This means that the activity is likely to be strictly defined by the teacher or by the program being used, allowing the child little opportunity to direct her own learning.

As both teachers and pupils become increasingly familiar with the presence of the computer in their classroom, the concept of "fair turns" for the motivation and entertainment of the children becomes less appropriate. With the opportunity to observe groups and to discuss and reflect on lessons, teachers gained greater insight into individual differences between children as described above and became more conscious that different children may actually need different experiences, some of which may be more appropriately met in non-computer activities. Their planning, which had been centred on the technology - taking a particular piece of software as the starting point and developing objectives for their lessons from it - became more centred on the needs of individuals and groups, starting with a mathematical topic and their knowledge of their pupils and deciding how (and indeed whether) to make use of the software for different groups of children.

As well as planning differentiated activities to address different needs within the same classroom, management issues arise when several groups are engaged in more than one activity. In order to make most effective use of the teacher's time some of the groups need to be able to work independently while others have the teacher's attention. It is obviously desirable for the teacher to anticipate this in her planning. The issue of teachers' awareness of the need for intervention and techniques of intervention arising from their computer experience is discussed further below.

The question of how groups are formed within the class is also an area in which the use of the computer appears to have influence. Some teachers involved in the in-service programme commented on the difference that using the computer made to children's willingness to work with one another:

> . . . it's a natural way of drawing children together from different parts of the classroom

> They don't mind their friendship groups being split up . . . If they're on the computer they work better together than when they weren't. They did their own thing if they were just at the tables.

It is possible that children who do not usually choose to work together may be prepared to collaborate when using the computer and may subsequently share work away from the computer.

Assessment

One of the worries that is expressed by teachers just starting to use the computer in their classrooms is about the lack of a product to be assessed. The display on the screen is transitory and the achievement of successful solutions to problems is not recorded. This is also a characteristic of many other practical activities - another area where mathematics teachers are often wary to tread.

Several of the teachers involved in the in-service programme were initially very concerned about the lack of a written record of the children's work. Some even made structured worksheets for the children to complete while using the computer to ensure that their achievement could be monitored. This obviously had the effect of directing and restricting the children's activity, even though this may not have been the teachers' original intention. They soon found, however, that they were making use of 'incidental teacher evaluation' (Mehan, 1989) on a regular basis. The visibility of children's strategies when using the computer and the opportunity to work closely with small groups of children allowed the teachers to recognise that valuable insights into children's understanding could be achieved by observation and by talking to children about their work. Such observation also brought to light unexpected aspects of the children's understanding which would have been unlikely to be discovered using written records alone.

As well as changing attitudes towards assessment methods, the visibility of the computer screen feeds into planning as teacher P commented:

> I can plan which groups I need to focus on much better as I have a better idea of what the children can do and understand. Usually I choose which group to focus on more or less at random . . *[but]* . . You can look at the computer screens and see what they are doing.

Gradually, worksheets became far less in evidence and many of the teachers began to be involved in discussion of and experimentation with alternative methods of non-written classroom-based assessment, not only for computer related activities.

Teacher Control and Intervention

The issue of the teacher's influence on children's learning needs to be considered at two levels: curricular planning, both long term and day to day, and teacher/pupil interaction. The way in which using a computer may enable children to encounter mathematical ideas before they would formerly have been introduced has been addressed already. Both individual teachers and mathematics departments as a whole need to take this into account when planning schemes of work, not only possibly changing the order in which some topics are taught but also having a more flexible attitude to curricular planning. At a time when the introduction of the National Curriculum in England and Wales is making flexibility more difficult and is, moreover, apparently laying down the progression of topics by law, few mathematics departments may feel able to take this on board. At the classroom level, however, there is a tension between computer use and rigid curricular planning which must be resolved by individual teachers. With school-based support during the in-service programme, some moved away from providing structured worksheets to guide children's experiences and towards posing more open questions and allowing children more freedom to choose their own strategies and targets, in both computer and non-computer activities.

The importance of the effect of teacher/pupil interaction on children's learning becomes more apparent to teachers when they have the opportunity to observe and work closely with individuals and small groups of children. The organisation of group work associated with computer use and, even more importantly, the characteristic of visibility of computer activity mean that these opportunities arise perhaps more frequently when children are using computers than during other types of work. Teachers involved in team teaching during the in-service programme became extremely aware of the effects, both affective and cognitive, that their interventions could have on children. One of the most frequent comments made after nearly every lesson was "I wish I'd said . . . " or, even more often, "I wish I hadn't said . . . ". In many cases this was a sign of a growing awareness that too much teacher direction could prevent a child from making, possibly unpredictable, mathematical discoveries for herself. It also led to greater anticipation of possible areas of difficulty or opportunities for learning where intervention might be appropriate. At the end of the in-service programme, during a centre-based session, the teachers were asked to plan a lesson using a specified piece of software. Those who had been involved in team teaching were more able to identify in their planning where teacher intervention might be appropriate and also appeared to be more prepared to allow children time to explore their own ideas.

CONCLUSION

The changes in classrooms and in teachers' and pupils attitudes described in this chapter arose in conditions which provided support for the teachers involved and encouraged them to reflect on their own practice and on the learning taking place in their classrooms. Within such an environment, specific characteristics of computer use can raise issues which facilitate change in the mathematics classroom culture: in particular, a shift in the control of the mathematical activity away from the teacher and towards the pupil. In different conditions, however, a number of factors make such changes less likely to occur.

Technical and classroom management problems have discouraged many teachers from attempting to use computers in their lessons. Even where a teacher is motivated to try to introduce the computer, the unfamiliarity of the medium means that without support technical issues may get in the way of educational ones. Dealing with moving classes of children into the computer room, connecting machines, or programs which will not load becomes a more urgent priority than thinking about the mathematical activity that is going on in the classroom. In many schools it is still difficult for a mathematics teacher to have regular access to a single computer in her own classroom; this is a major barrier to change.

If computer activities are seen as different and separate from other mathematical activities within the classroom then the characteristics of computer use are unlikely to affect the classroom as a whole. Centralised computer laboratories not only make frequent access and the use of the computer as a resource nearly impossible but also create the impression for both teachers and pupils that they are "doing computers" rather than "doing mathematics". The often cited suggestion that it is good to use computers because they are motivating also diverts attention from considering whether they are good for doing mathematics; use of the computer becomes a game rather than a part of the child's mathematical experiences.

The existing classroom culture affects the way in which the computer is used when it is introduced; the computer will only have an effect on the classroom when it is seen to be a natural part of it rather than fetishised as "new technology". We cannot assume that there is going to be a "computer revolution" but, where the computer is integrated into the culture of the mathematics classroom, it may play a role in creating tensions which lead to the development of new relationships between teachers, children and mathematics.

REFERENCES

Ball, D.: 1990, What is the role of IT within the National Mathematics Curriculum?, *Journal of Computer Assisted Learning* 6(4), 239-245.

Fey, J.T.: 1989, Technology and Mathematics Education: a Survey of Recent Developments and Important Problems, *Educational Studies in Mathematics* 20, 237-272.

Goldstein, R.: 1990, Adventure games don't rule, OK?, *Micromath* 6(2), 23.

Hawkridge, D.: 1983, *New Information Technology in Education*, London:Croom Helm.

Hillel, J., Kieran, C. and Gurtner, J-L.: 1989, Solving Structured Geometric Tasks: The Role of Feedback in Generating Strategies, *Educational Studies in Mathematics* 20(1), 1-39.

Hoyles, C. and Noss, R.: 1987, Synthesizing mathematical conceptions and their formalisation through the construction of a Logo-based school mathematics curriculum, *International Journal of Mathematical Education in Science and Technology* 18(4), 581-595.

Hoyles, C. and Sutherland, R.: 1989, *Logo Mathematics in the Classroom*, London:Routledge.

Lemerise, T.: 1990, Integrating Logo in the regular mathematics curriculum. A developmental risk or opportunity, *Proceedings of Fourteenth Meeting of International Group for the Psychology of Mathematics Education, Oaxtepec, Mexico*

Mehan, H.: 1989, Microcomputers in Classrooms: Educational Technology or Social Practice, *Anthropology and Education Quarterly* 20, 4-22.

Morgan, C.: 1990, Carry on SMILEing, *Micromath* 6(3), 14-15.

Morgan, C.: 1991, In-service Education, Computers and Action Research, *Paper presented at the Working Group on 'Teachers as Researchers' at PME 15*

National Curriculum Council: 1989, *Mathematics in the National Curriculum: Non-Statutory Guidance*, HMSO, London.

Noss, R.: 1988, The Computer as a Cultural Influence in Mathematical Learning, *Educational Studies in Mathematics* 19, 251-268.

Olson, J.: 1980, Teacher constructs and curriculum change, *Journal of Curriculum Studies* 12(1), 1-11.

Olson, J.: 1988, *Schoolworlds/ Microworlds: Computers and the Culture of the Classroom*, Oxford:Pergamon Press.

Papert, S.: 1980, *Mindstorms: Children, Computers, and Powerful Ideas*, Brighton:Harvester Press.

Taylor, R.: 1980, *The Computer in the School: Tutor, Tool, Tutee*, New York:Teachers College Press.

Watson, D. M.: 1990, The Classroom vs the Computer Room, *Computers Education* 15(1-3), 33-37.

Weir, S.: 1987 *Cultivating Minds: A Logo Casebook*, London:Harper and Row.

Wright, A.: 1987, The Process of Microtechnological Innovation in Two Primary Schools: a case study of teachers' thinking, *Educational Review* 39(2), 107-115.

Candia Morgan,
Department of Mathematics, Statistics
and Computing,
Institute of Education,
20, Bedford Way,
London WC1H 0AL U.K.

D. PIMM

SPOKEN MATHEMATICAL CLASSROOM CULTURE: ARTIFICE AND ARTIFICIALITY

INTRODUCTION

"I'll offer a trade", said Nellie.
"A trade?" Kate had reached, she realized, the moment in certain conversations when she found herself merely echoing the words of her interlocutor. It was evidence of unhappiness. (Cross, 1991, p. 134)

One major component of an educational researcher's task is to document and understand the world of schools, teaching and learning: in the case of mathematics education researchers, that of mathematics classrooms. Using metaphors is one way of undertaking this complex task. For instance, at the most general level, when looking at discourse generated about teaching, both academically and popularly, a number of such metaphors can be seen at work, each with its own imagery and accompanying rhetoric. Examples include seeing teaching as gardening (kindergarten, growth, nurturing, but what about weeds?), as doctoring (remediation, diagnosis, prescription, but misconception as illness?), as coaching (basic skills, exercises, repetition, competition and performance) and as acting (classroom roles, stage-managing events, teacher as director or actor, pupils as actors or 'audience').

Metaphors act to highlight possibly unnoticed aspects of the activity as well as suppressing others according to the particular perception offered by the invitation to see one thing as another. Lakoff and Johnson (1980) in their book 'Metaphors We Live By', explore systematic metaphors in English that are in common use. In this chapter, I want to explore another metaphor, one prompted by a quotation educational linguist Michael Stubbs: "There is a sense in which, in our culture, teaching *is* talking" (1983, my emphasis). Reading this statement as a metaphor affords a perception on teaching and encourages the question: Is it any kind of talking which is teaching, or are there particular kinds?

How might we explore this question? An over-discussed split in educational research is betwen quantitative and qualitative methodologies. (For more discussion on this ostensible dichotomy of aims and approaches in mathematics education, see Scott-Hodgetts, 1992.) One intent of quantitative methodologies is to provide a picture composed of many responses, usually of the 'same' thing (though recall Heraclitus' observation concerning our inability to step into the same river twice, or a later concern with stepping into the 'same' river at the 'same' time, yet experiencing it differently). A

S. Lerman (ed.), Cultural Perspectives on the Mathematics Classroom, 133–147.

quantitative approach may attempt to provide a picture or model of, for instance, 'an ideal native informant' or 'a typical classroom interaction'. Walkerdine (1988, 1990) has fundamentally criticized at length the psychologists' creation of 'the child'.

Qualitative methodologies attempt to provide us with an insider's perspective: a qualitative approach can inform us of the richness and particularity of an actual classroom. Stephen Brown has commented: "One incident with one child, seen in all its richness, frequently has more to convey to us than a thousand replications of an experiment conducted with hundreds of children" (1981, p. 11). Yet do we not only find out about the particularities and contingencies of that classroom? One danger with videotaping classrooms in particular is to succumb to the feeling that you can see 'reality' transparently and that you can consequently know all there is to know about these settings and their participants.

A general concern arising from the foregoing involves aspects of representativeness and voice - who gets to speak? Whom do we hear from, and how? One result of the disciplined ways of looking that many fields develop may be that the same voice (or voices) gets replicated over and over. (See, for example, Belenky *et al.* (1986) or Spivak's (1988) paper entitled*Can the subaltern speak?*) In this chapter, I want to take this very general concern from ethnographic research, namely 'Who gets to speak, how and about what?', and ask it of mathematics classrooms.

But my focus here is not on questions of representativeness of various ethnic or social groups (important though such questions can be, and are explored in some of the other chapters in this volume), but more on the form and structure of spoken interations between mathematics teachers and pupils. In particular, I am interested in the control and shaping of the discourse in a mathematics classroom, arguably in many instances towards the teacher's own ends. The classroom transcripts I offer should be seen as illustrative rather than generative of the categories I use and questions I pose, though in some ways, this too is an artificial distinction, as these categories have arisen through looking at many hours of videotape of actual classrooms.

In *Speaking Mathematically* (Pimm, 1987), I introduced the metaphor of a teaching 'gambit'. One feature of a gambit in chess is that there is some sacrifice involved in order to move toward a hoped-for, long-term advantage. I exemplified the notion of teaching gambits by three situations: having pupils talk in pairs, not answering a question directly, and the use of silence. I then went on to look at common ways of asking, answering and deflecting questions in mathematics classrooms, with a tacit focus on what some of the sacrifices might be as well as the hoped-for advantages.

In this chapter, I want to pursue similar themes, specifically those of teacher gambit and intention as embodied in particular discursive practices. My focus is on specific aspects of spoken classroom interactions between teachers and pupils and their significance, highlighting in certain circumstances the (to me) interesting or incongruous usages that pertain in classrooms, particularly when mathematics is under discussion. The general frame will involve the notion of meta-commenting, though first I look closely at the situation of whole-class report-backs, an increasingly common feature not only of 'investigation lessons', but many mathematics lessons in general.

There are many different aspects of and relationships between mathematics and language that can be highlighted as part of the mathematics education enterprise. My intention in this chapter is to start to explore some of the particularities of teacher - pupil spoken interactions, but I feel that it is of crucial importance to find ways of talking about the varied components of mathematical activity itself. One means for achieving this would be to focus attention on particular features of doing mathematics which might then afford teachers greater insight into what is happening in and between their pupils when in the mathematics classroom. This chapter, and indeed this volume, focuses on aspects of the culture of doing and teaching mathematics, in particular on the social contexts and communities which classrooms exemplify.

CONTROLLING REPORTING BACK TO THE WHOLE CLASS

I look first at the situation of report-backs in classrooms and some of the linguistic demands this places on teachers and pupils and discuss some ways in which these demands are met in practice. The 'investigation lesson' in mathematics has its common structure of whole-group posing of the task, pupils then working together in small groups and finally some form of reporting back to the whole class. The activity of reporting back throws up some interesting linguistic questions, in relation to teacher and pupil language and styles of speech in the context of the mathematics classroom. One key question for me is: 'For whose benefit is the reporting back?'. There seem to me to be at least three possible answers and corresponding sets of potential justifications.

(a) The pupil(s) doing the reporting: for them, plausible justifications include development of a range of communication skills, their use of language and development of social confidence. A further important possibility is that of becoming skilled at reflection on a mathematical experience, and distilling it into a form whereby they as well as others may learn from it. What sort of language (style, structure, organisation, register, and so on) does the reporter use?

(b) The other pupils listening: for them, potential benefits include hearing alternative approaches and that perhaps others besides themselves had difficulties. In addition, they have the chance to ask genuine questions of other pupils and to engage in trying to understand what others have done on a task that they also have worked at.

(c) The teacher: potential benefits here include a range of opportunities to make contextually-based meta-remarks about methods, results and processes (perhaps indicating the task is open and that a number of ways of proceeding are possible or different emphases can be placed), as well as to value pupils' work and broaden pupils' experience (possibly by offering an original idea to the rest of the class). "Did anyone else ..."

Reporting back can also provide the teacher with access to what the pupils think the task was about, as well as less tangible but nonetheless interesting information such as what they think is important (in terms of what they select to talk about) and their level and detail of oral expression of mathematical ideas. But it can also be *assumed* that reporting back *has* to happen, perhaps meeting the need (or expectation) for some sort of whole-group ending, 'pulling the session together'. If that is the case, then many of the above benefits may not necessarily accrue, because attention is not being paid to their elicitation.

Here are three further questions that, for me, encapsulate some of the key issues for the teacher involving mathematics education and language arising from the task of reporting back on mathematical investigations. (For further analysis of this particular situation, see Pimm, 1992.)

(1) *How can one contend with the tension between wanting the pupil(s) to say themselves what they have done, while wanting to use what they say to make general remarks about how to undertake investigative work?*

This can be particularly strongly felt in the case where the teacher has seen something that she feels is an instance of a higher-order process that is valued (be it specialising systematically, developing notation, coping with getting stuck, or whatever) while circulating around the small groups. When invited to tell the class about this incident, it may well have not been a salient one for the pupil(s) (unless the teacher made a big point of it at the time), so they probably have little idea either of what to emphasise or why this particular incident is being focused on. And if the teacher has made the point to them already, why are they being asked to repeat it?

There seems to be a reluctance for teachers to say for them what one of the pupils has done, instead of prompting with an explicit comment something like: "Why don't you say it, because you did it?" A truthful response from the pupil might be: "Because it is *your* anecdote. I don't know what significance it has *for you*". They do not have the teacher's reasons for highlighting particular parts of what they have done.

(2) *How can pupils develop the linguistic skills of reflection and selection of what to report? How can they work on acquiring a sense of audience?*

Obviously, this is contingent on the perceived audience and purpose behind carrying out the reporting back. Who knows what might be worth telling about? The teacher has no control over what comes out. She can only go to work on the utterances *after* they have emerged from the pupil, though it is likely that what she then does with the response will influence later reporters.

One important ability of pupils is to be able to disembed their discourse from the knowledge of the group who saw the work developed, in order that someone who was not there can follow what is being described. The tendency is for the reporter to assume that everyone will know what they are talking about.

(3) *To whom is the reporter talking?*
Pupils in such a situation often address themselves to the teacher, the person who, besides their own group, probably knows *most* about what they have done - and the other pupils know this. If the teacher reinterprets what the pupil says for the rest of the class (possibly by playing a role which in television interviews is known as 'audience's friend'), what effects might this have on the reporter? By playing 'audience's friend', the teacher can also ease the strain on the reporter, by taking the focus off them, possibly by reinterpreting or expanding for the audience and then moving into more general questioning of the class.

However, it also may result in the teacher being looked to as broadcaster and interpreter of the person reporting back and thus acting as an intermediary between reporter and audience which may well get in the way of direct communication. After all, if reporting back were an effective technique in and of itself, there would be little need for the teacher to intervene - the reporting back would do its own work.

Where is the teacher standing and where is the control? If there is a silence, whose responsibility is it to fill it? Where is the audience's attention and who are they asking questions of? If it is predominantly a conversation between the reporter and the teacher, what is the intended role for the pupils who are being invited to listen?

In this section, one focus was on how the teacher can apparently invite direct communication with the class from a small group of pupils, while retaining his or her teaching agenda. One link with the subsequent section is that occasions where the teacher acts as 'audience's friend' require language apparently in the conversation yet about the conversation itself - the making of such meta-remarks by teachers providing the subsequent, central focus of this chapter.

SOME CLASSROOM DISCOURSE PATTERNS

Since the mid-1970s, discourse analysis has been used to explore aspects of classroom discourse, among other linguistic settings, and to highlight certain normative aspects of language use in these particular speech settings. One early 'finding' by Sinclair and Coulthard (1975) was the almost incessant repetition of the sequence I(nitiation) - R(esponse) - F(eedback) in teacher-pupil exchanges. (A more detailed account of this sequence and some transcripts from lessons where teachers have found ways of escaping it is given in Pimm, 1987.) One sense of unease that undertaking discourse analysis has engendered is its necessary ignoring of the content of the discourse, of what is being taught, in favour of its form. This is a consequence of the nature of the actual enterprise.

Further comments about what discourse analysis cannot offer are made by Edwards and Mercer (1987) in their book *Common Knowledge*. They comment (p. 10): "It may be thought that a concern with the content of the talk rather than with its form, and with interpreting people's meanings rather than coding their turns at speaking, is an altogether less rigorous and objective way of dealing with discourse", but then go on to offer three justifications for so doing (formal discourse analysis doesn't allow them to answer the questions they want to ask; their analyses are offered in terms of the data themselves, not data already coded; discourse analysis itself also needs an interpretative framework in order to make judgements about coding). As we shall see in this chapter, my intent here is to look at situations where *teachers* opt to ignore content in favour of form as part of their teaching discourse.

Edwards and Mercer examine the rhetoric of 'progressive' education in English primary schools and focus on the disparity between the level of freedom accorded the pupils at the level of action and that at the level of discourse and 'generation' of the knowledge. They also detail various indirect teacher devices for constructing the 'common knowledge' in the classroom, identifing: controlling the flow of conversation; determining who is allowed to speak, when and about what; use of silence to mark non-acceptance of a pupil's offering; reconstructing (and reformulating) 'recaps' of what has been said, done or ostensibly discovered.

One key focus they allude to is that of teacher questions (p. 30): "Teachers may all be obliged to control classes and lessons, but they choose particular strategies for doing so. ... However, there has been hardly any research on teachers' purposes in asking questions". One of those who have undertaken this topic is Janet Ainley (1987, 1988), who has explored the varied functions of questions and how they are interpreted by pupils. With any question there is the notion of level. If a teacher arrives at a table asking "What question am I about to ask you?", a different interpretative task is being offered from the one initiated by the request "Tell me what you are doing".

In what follows, I look at the notion of meta-commenting as a systematic feature of mathematics classroom discourse.

META-COMMENTING

In *Up the Down Staircase*, novelist Bel Kaufman (1964) tells poignantly of a male English teacher receiving a love letter from a female pupil. The way he contends with this is by correcting the spelling and grammar, making some remarks about clichéd use of language, and then returning it to her with a grade. Such an extreme instance of attending to form over content merely serves to highlight one important possibility for unusual discourse in classrooms: precisely because his actions would have been entirely unexceptional and appropriate *within* a clssroom setting.

In an paper entitled 'Organizing classroom talk', Stubbs (1975) offers the notion that one of the characterising aspects of teaching discourse as a speech event is that it is constantly organised by meta-comments, namely that the utterances made by pupils are seen as appropriate items for comment themselves and, in addition, that many of the meta-remarks are evaluative. He comments:

> The phenomenon that I have discussed here under the label of meta-communication, has also been pointed out by Garfinkel and Sachs (1970). They talk of "formulating" a conversation as a feature of that conversation.
>
> > "A member may treat some part of the conversation as an occasion to describe that conversation, to explain it, or characterise it, or explicate, or translate, or summarise, or furnish the gist of it, or take note of its accordance with rules, or remark on its departure from rules. That is to say, a member may use some part of the conversation as an occasion to *formulate* the conversation."
>
> I have given examples of these different kinds of "formulating" in teacher-talk. However, Garfinkel and Sacks go on to point out that to explicitly describe what one is about in a conversation, during that conversation, is generally regarded as boring, incongruous, inappropriate, pedantic, devious, etc. But in teacher-talk, "formulating" is appropriate; features of speech do provide occasions for stories worth the telling. I have shown that teachers do regard as matters for competent remarks such matters as: the fact that somebody is speaking, the fact that another can hear, and whether another can understand. (pp. 23-4)

In Stubbs' paper itself, some of the examples he uses to exemplify his various types of speech acts undertaken by teachers are taken from a lesson involving an English language teacher working with a group of teenage pupils who are non-native speakers. Nonetheless, despite the fact that in such a second language class matters of language and the form of utterances are obvious foci for attention, I am taken with his ostensible characterisation (in part) of teaching in terms of making meta-comments. I begin with a straightforward example from a mathematics lesson.

In a class of nine-year-olds, the teacher has two pupils reporting back to the class on their table of results.

T: You can tell us what *these* numbers are, these ones ... and then you can tell us what these ones are.
P: These are the number of people and these are the number of handshakes.
T: Right, did everyone hear them?
[Chorus of yesses and nos.]
T: Right, say it again Stuart.
Stuart: These are the numbers of people and these are the number of handshakes.

Such an interchange seems entirely fit and proper: the monitoring of what is going on and whether people can hear is part of being a teacher. In this class, though, the teacher frequently 'echoes' her pupils' contributions (recall the opening quotation), but in a 'rebroadcasting' manner (as mentioned earlier in the context of report-backs), making the observation aloud that Sara has just said this or Winston that.

T: Ten times a thousand. What's the next one going to be? Dudu?
Du: A million
T: She says it's going to be a million. How many noughts in a million?
Da. Four.
T: A million has four noughts, Danny says. Who said millions? Was it you, Emira?

But as well as ensuring that the others in the class have heard what has been said here (this particular class is lively and noisy generally), saying back to the class what someone has just said (to her, given the customary directionality of the classroom channels of communication) also allows her to avoid giving overt feedback, while still making an utterance, while still taking a turn in the conversation.

T: Shiral says they're the eight times table. Ben?
B: They're all multiples of six.
T: Ben says they're all multiples of six. How do you know they're all multiples of six?
B: Because they're even, ... and so they can all be done. And also ... it's hard to explain.

It allows her to put pupil proposals into opposition without herself having to decide among them. But her spoken style makes it particularly crucial that her pupils be attuned to those occasions when a teacher comment is intended for the whole class and those when it is focused on a particular pupil.

Consider now the following excerpt from a class of twelve-year-olds.

T: Uh-huh. Thank you very much. I've got another one. [Writes on board.] But I'm not too sure whether I have done it right or not. Have I?
Pupils: Yes, no, no. [More nos.]
T: It's not quite right - OK, could someone make a change so that it is right. ... Thanks, Zena.
Zena: Can I just rub it out?
T: Yes, do. [With slight irony, as she has already rubbed out the final 3 with her finger and changed it to a 4.] You can even use a board rubber if you want to.
Zena: [Looks at Dave who is standing at the back of the class] Is that all right?
Pause (2 secs)
T: Zena asked a question.
[Chorus of yesses, the lesson continues.]

I am particularly interested in the last interchanges where Zena asked two questions. She may not have distinguished between them, yet they received quite different treatment from the teacher. The first one was a request for confirmation of procedure and permission - which was given - and the second was a request for verification, which evoked a meta-comment to the class to the effect that she had asked a question.

What sense can be made of the teacher commenting on something that everyone already knows, at least if they were attending to Zena at the front of the class? The teacher had ostensibly made a response, taken a turn in the conversation as the question had been addressed to him. But what he chose to do was to draw the class' attention to the form of the pupil's utterance as his response to it, rather than responding on the level at which it was made. The teacher himself had used those exact same words ("Is that alright?") a few moments earlier: indeed, such a request for audience confirmation was built into the activity.

Below I give a short extract from a secondary mathematics class in which I feel something similar occurs. David Cain (DC) is working with a group of fourteen-year-olds and the task he has set them is to go from the net of a cube, with which they are familiar to one for a solid (which he terms a 'slanty-cube') where each face is a rhombus. The lesson extract opens with the word 'rhombus' written on the board and the net of a cube drawn. He says to the whole class:

> Therefore, all you've got to do is this - very simple. There's the net of a cube - you've got to make these into rhombuses and it should stick together to make a ... slanty-cube. Yes, and instead of a cube, you'll have what is called a rhomboid - right?

Later in the lesson, he is talking with one particular girl, G, about why her paper model does not work. All of the language is quite inexplicit as both of their attentions are focused on the paper model throughout the conversation. [I have tried to give some feel for the overlapping nature of the conversation and also the relative duration of pauses. '...' means a short pause (up to half a second), longer pauses are marked with approximate durations.]

DC: This should fold up there.
G: No, it doesn't
DC: But it don't, does it.
DC: Why, why doesn't it?
G That should be tilted that way.
DC: *That* one, *that* way. [checking] And then?
G: That way, and ... no, no. [Turns over piece of paper a couple of times. trying to see how it should go.]
Pause of several seconds
DC: [Laughs.]
G: [Laughs slightly ruefully.] If I tilt them top three and then that bit ...
DC: Right.
G: ... would that work?
Pause (2 secs)
(**)
DC: What do you mean, "Would that work?"? You're asking me. What do you think? You think I'm going to tell you?
G: No [laughs].
DC: No, all right then. [DC takes model]
Pause of seven seconds, both looking at model.
DC: [pointing at one particular fold] Is that alright?
(**)
G: mm-hmm [affirmative]
DC: Now what's the problem now - with that one?
G: That - that should be onto there.
DC: Yup - but it doesn't does it - you get a gap. OK, so what have you got to do?
G: Move that one to there.
DC: OK, you can try that. ...

I am very interested in the exchange marked between the sets of asterisks. David has not answered her question, in keeping with his desire to have pupils validate their work when they are able to. He does this by drawing her attention both to the question she has asked (by repeating it) and then by asking her whether she thinks it likely that he is going to answer it.

One interesting question is once you have raised the discussion to a meta-level, how do you then return it to the normal one where the form of utterances is not uppermost as the topic of attention. In this instance, David achieved this by means of a long pause and then offering a vary particular focus of attention back on one fold of the model and the question "Does this work?". Interestingly, the pupil did not turn the teaching gambit on its head by mirroring his earlier response and replying "What do you think? You think I am going to tell you?". This provides a nice instance of different rules applying to the teacher and the pupils - making meta-comments on the discourse is the prerogative of the teacher.

Moves to and from a meta-level need to be handled smoothly if they are not to be too disruptive of attention and focus, yet not so smooth as to pass unnoticed, as they are often important sites for teaching and, hence, potential learning. The teacher question cited earlier "What question am I about to ask you?" is a meta-question designed to encourage pupils to notice the teacher's interventions as regular and systematic. It also carries with it the suggestion that the pupil might take on the function that the teacher has been carrying out up until now by asking the same question of herself.

Eric Love has commented that as a teacher he often asked: "What do you want me to tell you?" in response to a particular request from a pupil for help, sometimes going further to indicate a range of possible things he was willing to answer. ("I could tell you exactly the answer, though that might not help you next time you have a similar problem. I could show you how to do this one, show you another one like it, ...".) In so doing, he has drawn attention to the fact of the situation as he sees it - that *I* (as teacher) am here to help *you* (as pupil) to learn *mathematics* - but that help might take a variety of forms. (Recall also the earlier brief comments on teacher questioning and level.)

This returning choice to the pupils may seem a preferable solution to the problem of apparently ignoring or going against direct requests for information or help, which if they are not going to be responded to exactly at face value, require some means or gambit for their deflection. This presents a teaching tension. If the teacher only makes meta-remarks, only answers questions with questions, the pupil may well lose confidence or trust in the teacher as a source of a conversation about the *content* they are struggling with. Yet, if the teacher only engages with the content, this in part rescinds the possibility of the teacher teaching.

This seems to be a related to the following formulation of the didactic tension (Mason, 1988).

> The *more* explicit I am about the behaviour I wish my pupils to display, the more likely it is that they will display that behaviour without recourse to the understanding which the behaviour is meant to indicate; that is, the more they will take the *form* for the substance.

The less explicit I am about my aims and expectations about the behaviour I wish my pupils to display, the less likely they are to see the point, to encounter what was intended, or to realise what it was all about.

Teaching necessarily operates within the constraints of this tension, and the phenomenon of meta-communication provides one instance of how it is lived by teachers in practice.

Rosemary Clarke (1988), in her comparison of teaching and gestalt therapy, draws attention to the role of the therapist in attending to the process her client is engaged in by attending to the content - but not getting caught up in that content. One feature of this can be indirectly to encourage reflection on the possible nature of the speech act just made. In a similar way, the teacher must be able to stand outside the discourse as a commenter on it in order to teach, yet still be seen as a participant in it.

I am not proposing 'teaching as therapy' as another metaphor. Rather, I am suggesting that psychotherapy involves one particularly pure form of teaching, one where the form of what is said is highly in focus by the therapist, who customarily operates with a working belief that what is said *is* what is always meant.

To repeat myself, if teacher-talk is only at the level of meta-remarks, the pupil may lose confidence in the teacher as a source of information about the *content* they are struggling with. They may well come to the conclusion that the teacher is not being straightforward with them. Yet, if the teacher only engages with the content, there is the difficulty of 'teacher lust' (to use Mary Boole's evocative term - see ATM, 1980, p. 11), of the desire to tell the pupil things[1], which in part negates the possibility of the teacher teaching. It is in this area of tension that the artifice of teachers and the artificiality of teaching discourse appear.

---

[1] I do not mean to suggest in the slightest that a teacher should never tell a pupil things. This belief is one of a number of simplistic contemporary tenets that seem to me foohardy. But it is prcisely when and what things that any individual teacher would tell that makes the issue an interesting one. See Smith (1986) and Ainley's (1987) rebuttal, as well as Brown (1991) and, on the topic of the teacher's *desire* to tell the pupil things, Tahta (1991).

CONCLUSION

Reporting back can place some quite sophisticated linguistic demands on the pupils in terms of communicative competence - that is, knowing how to use language to communicate in certain circumstances. Here, it includes how to choose what to say, taking into account what you know and what you believe your audience knows. Stubbs claims (1980, p. 115): "A general principle in teaching any kind of communicative competence, spoken or written, is that the speaking, listening, writing or reading should have some genuine communicative purpose". Yet this is at odds with my view of the classroom as an avowedly, deliberately un-natural, artificial setting, one in which the structure and organisation of the discourse by the teacher has some quite unusual features. The artifice of the teacher is in rendering them sufficiently unexceptionable that the conversations that occur can flow smoothly and effectively.

Pupils learning mathematics in school in part are attempting to acquire communicative competence in mathematical language, and classroom activities can be usefully examined from this perspective in order to see what opportunities they are offering pupils for learning. Teachers cannot make pupils learn - at best, they can provide well-thought out situations which provide opportunities for pupils to engage with mathematical ideas and develop skills in using spoken and written language to that end.

Teachers, in order to teach, need to develop or acquire linguistic strategies (which I earlier called 'gambits') in order to, among other things, direct pupil attention to salient aspects of the discourse - or indeed the nature of that discourse itself - while still remaining in 'normal' communication with the pupil. In focusing on instances of meta-commenting, I have tried to highlight specific clssroom incidents which seem to me central to the teaching enterprise as a whole.

The asymmetry of power between teacher and pupil is visible in the discourse structures. As Foucault reminds us, power and its ramifications are always with us, not least in educational settings. It does not seem to me inappropriate that teachers should be powerful individuals, nor that they should use their positional power as teachers in order to teach. I am curious about how different teachers manifest this control through their discourse, and the similarity of ways that recur to achieve certain functional ends. As one teacher has commented, "the teacher always teaches" (Jaworski, 1991). I am also continuingly curious about how teachers come to teach as they do, how they overtly learn or otherwise acquire the relevant discourse patterns, ones which can on occasion seem so peculiar by general social norms for conversational style and structure.

NOTES

I am grateful to Barbara Jaworski and Eric Love for conversations on different parts of this chapter.

The transcripts provided were taken from videotapes made in the course of preparation of the two Open University mathematics education courses ME234 *Using Mathematical Thinking* and EM236 *Learning and Teaching Mathematics*. Further information about the availability of these tapes is available by writing to the Centre Secretary at the address given for the author.

REFERENCES

Ainley, J.: 1987, Telling Questions. *Mathematics Teaching*, 118, pp. 24-26.

Ainley, J.: 1988, Perceptions of teachers' questioning styles. *Proceedings of the Twelfth Conference of the International Group for the Psychology of Mathematics Education* Veszprem, Hungary, pp. 92-9.

ATM: 1980, *A Boolean Anthology*. Derby: Association of Teachers of Mathematics.

Belenky, M., Clinchy, B., Goldberger, N. and Taroule, J.: 1986, *Womens' Ways of Knowing: The Development of Self, Voice and Mind*. New York: Basic Books.

Brown, L.: 1991, Stewing in your own juice. In Pimm, D. and Love, E. (Eds.) *Teaching and Learning School Mathematics*. Sevenoaks:Hodder and Stoughton, pp. 3-15.

Brown, S.: 1981, Sharon's Kye. *Mathematics Teaching*, 94, pp. 11-17.

Clarke, R.: 1988, Gestalt therapy, educational processes and personal development. In Pimm D. (Ed.) *Mathematics, Teachers and Children*. Sevenoaks: Hodder and Stoughton, pp. 170-182.

Cross, A.: 1991, *The Players Come Again*. New York: Ballantine.

Edwards, D. and Mercer, N.: 1988 *Common Knowledge*. London: Methuen.

Garfinkel, H. and Sacks, H.: 1970, On formal structures of practical action. In McKinney, C. and Tiryakian, E. (Eds.), *Theoretical Sociology*. New York:Appleton Century Crofts, pp. 337-366.

Jaworski, B.: 1985, A poster lesson. *Mathematics Teaching*, 113, pp. 4-5.

Jaworski, B.: 1991, Interpretations of a Constructivist Philosophy in Mathematics Teaching Unpublished Ph.D. thesis, University of Birmingham, Birmingham.

Kaufman, B.: 1964, *Up the Down Staircase*. New York: Avon.

Lakoff, G and Johnson M.: 1980, *Metaphors we live by* Chicago:University of Chicago Press.

Mason, J.: 1988 *What to Do When You are Stuck*. (ME234 *Using Mathematical Thinking*, Unit 3). Milton Keynes, The Open University.

Pimm, D.: 1987, *Speaking Mathematically: Communication in Mathematics Classrooms*. London: Routledge.

Pimm, D.: 1992, 'Why are we doing this?' Reporting back on mathematical investigations. In D. Sawada (Ed.), *Communication in Learning Mathematics*. MCATA: Edmonton, Alberta.

Scott-Hodgetts, R.: 1992, *Learning* Mathematics: Pupil Perspectives. (Monograph L, EM236 *Learning and Teaching Mathematics*). Milton Keynes: The Open University.

Sinclair, J. and Coulthard, M.: 1975, *Towards an Analysis of Discourse*. Oxford: Oxford University Press.
Smith, J.: 1986, Questioning questioning. *Mathematics Teaching*, 115, p. 47.
Spivak, G. C.: 1988, Can the subaltern speak? In Nelson, C. and Grossberg, L. (Eds.) *Marxism and the Interpretation of Culture*. Chicago: University of Illinois Press, pp. 271-313.
Stubbs, M.: 1975, *Organizing classroom talk*. Occasional paper 19, Centre for Research in the Educational Sciences, University of Edinburgh.
Stubbs, M.: 1980, *Language and Literacy*. London: Routledge and Kegan Paul.
Stubbs, M.: 1983, *Language, Schools and Classrooms*. London: Methuen.
Tahta, D.: 1991, Understanding and desire. In Pimm, D. and Love, E., (Eds.) *Teaching and Learning School Mathematics*. Sevenoaks: Hodder and Stoughton, pp. 220-246.
Walkerdine, V.: 1988, *The Mastery of Reason*. London: Routledge.
Walkerdine, V.: 1990, Difference, cognition and mathematics education. *For the Learning of Mathematics*, 10(3), pp. 51-6.

David Pimm,
Centre for Mathematics Education,
The Open University,
Walton Hall,
Milton Keynes,
BUCKS MK7 6AA

Sinclair, J. and Coulthard, M.: 1975, *Towards an Analysis of Discourse*, Oxford University Press.

[illegible]

[illegible]

[illegible]

T. WOOD

PATTERNS OF INTERACTION AND THE CULTURE OF MATHEMATICS CLASSROOMS

Recent opinions of those interested in mathematics education have suggested that variations in the classroom culture and the nature of the patterns of interaction that occur between the teacher and the students create quite different settings for enhancing learning. Even earlier studies had pointed to the discrepancies between the intentions of teachers, their actual practice, and the latent learning of students (Bauersfeld, 1980; Holt, 1982; & Willis, 1977). In many ways, teachers have often unwittingly undermined their own goals by failing to realize that the consequences of their interaction are often quite different from their intentions. Descriptions of various mathematics classes commonly reveal them to be settings in which teachers still view their role as being responsible for only ensuring students' learn specified procedures for solving mathematical problems (Goodlad, 1983; Stodolsky, 1988). From this philosophy a long tradition of rule-bound school mathematics has continued. It has restrained the motivation for inventiveness for any student who enjoys both independent reasoning and collaboration of ideas. Children who engage in mathematical activity in interactive situations do learn mathematics in a manner that extends beyond the realms of memorized procedures.

From our past research in classrooms, we have also contended that distinct differences do exist between the nature of the traditional mathematics class and the culture of an inquiry class (Cobb, Wood, Yackel, & McNeal, 1992; Wood, Cobb, & Yackel, 1991). Micro-qualitative analyses of children's social interaction (Yackel, Cobb, & Wood, 1991) and their mathematical activity have suggested that the community that is constituted in the inquiry classroom creates possibilities for understanding mathematics that are not found in textbook-based classrooms. This assertion has been supported by the results from quantitative analyses of assessments of our primary classrooms. To this end, it has been found that children in classrooms which encourage collaborative interaction to solve challenging problems learn mathematics with greater understanding than do children in traditional classes (Cobb, Wood, Yackel, Nicholls, Wheatley, Trigatti, & Perlwitz, 1991; Wood & Sellers, 1993). Moreover, these children hold different beliefs about their role as students in these classes. They believe that it is their responsibility to make sense of mathematics and to make it understandable to others. In the past, although the teacher and the nature of teaching in these two settings has been taken into consideration somewhat, the major focus has been on students' learning. Thus, analyses and an

S. Lerman (ed.), Cultural Perspectives on the Mathematics Classroom, 149–168.

understanding of the teacher's activity in these inquiry classes has not been specified in as much detail as has that of the children's.

The intention of this chapter, then, is to deepen an understanding of the relationship between teaching and children's learning by examining patterns of interaction that are constituted in mathematics classes and the nature of the reasoning that occurs during this activity. The analysis of this discourse will attempt to move beyond previous considerations of form (e.g., Mehan, 1979; Stubbs, 1974) to examine the relationship between interaction and teaching and learning. I will first describe two distinct patterns of interaction that have been identified in mathematics classes and discuss the differences in the underlying theoretical orientation for each. An investigation of these patterns will explain the dynamic interactive processes by which mathematical meaning is constructed. Next, I will extend the discussion to consider the manner in which the differences in theoretical orientation create different cultures in classrooms which influence students' growth in mathematical understanding. Finally, I will present a fundamental limitation of considering Vygotskian or the neo-Vygotskian theory as the basis for improving the quality of teaching and learning mathematics in school classrooms. Central to this particular point will be the contention that differences in the cultures of mathematics classrooms are realized in the existing patterns of social interaction among the participants. In this regard, the culture established between the class members not only influences their mathematical understanding but also plays a major role in students' motivation for and beliefs about learning mathematics. (Bauersfeld, Krummheuer, & Voigt, 1988; Cobb, Wood, & Yackel, 1991, 1993; Cobb, Yackel, & Wood, 1989).

ROUTINES AND PATTERNS OF INTERACTION IN CLASSROOMS

A brief look at the research on classroom interaction reveals that much of the work either centers on examining the form and structure of the exchange between teacher and students or surveys the kinds of questions that are asked. In a majority of these studies, social interaction and communication are seen as means by which to regulate and constrain students' potential for learning (Erickson, 1982; Edwards and Mercer, 1987). Upon closer examination, we find the discourse takes on the form of the well known tripartite exchange. Briefly, this familiar discourse pattern consists of an initiation by the teacher, which is usually a question to which the answer is already known by the teacher, and a student response, followed by an evaluation from the teacher. This form of exchange is not unique to mathematics classes but is pervasive throughout all levels of schooling and has been fairly extensively examined (Mehan, 1979; Sinclair & Coulthard, 1976; Hoetker & Ahlbrandt, 1969).

Dissatisfied with the limitations of these results, some researchers have proceeded further to consider longer events (Mercer, 1992) and identify those patterns of interaction that occur repeatedly to form the regularity and predictability of classroom life (Voigt, 1985, 1990). These patterns reflect the nature of the interaction and the characteristics of the practices which have been mutually established among the individual members of the class. As such, they become the *taken-for-granted* ways of acting which form the smooth functioning of the class in which teachers and students jointly and interactively produce certain regularities in their actions and in their ways of communicating. An example of this is the ubiquitous "sharing time," described by Christie (1989) in which children are given time during class to tell about personal events to other members.

Initially, children adjust their actions as they learn to accommodate to the behavior of others and ultimately form consistent routines for acting in different classroom situations. These regularities that students develop enable them to maintain continuity in their daily activity without the necessity of deliberating among the various options for action. Consequently, the routines children develop allow many aspects of a situation to become predictable so that the focus of their attention can be on making sense of their activity when encountering unfamiliar or problematic aspects of a mathematical situation. These consistent and reproducible ways of behaving emerge for teachers as well as for students, and certain actions become reliably successful in coping with a situation. Most importantly, these individual routines evolve to form the interlocking ways of behaving that underlie interaction and the manner in which communication will occur in the classroom.

From the perspective of the observer, the patterns of interaction are seen as emerging from the individual's interpretation of anothers' actions and from the mutual orienting that occurs between the teacher and students. As they engage in a multitude of shared experiences in the classroom, they build up mutual, negotiated expectations and obligations over the course of the school year. This joint history acts to form the "collective intellect" (Bauersfeld, 1990) of the classroom, and with this a social equilibrium as individuals adjust their routines to achieve a sense of balance in which classroom events seem to flow in a seamless fashion.

The nature of the interaction that evolves among the members of the class reflects the mutually established network of obligations and expectations that form the basis for the taken-for-granted ways of acting. These social norms that are representative of the interlocking system of obligations and expectations that exist among the participants underlie the patterns of interaction that are necessary to reduce the complexity of the processes of classroom life. The teacher plays a major role in determining and guiding the development of these norms and in providing the means of support for maintaining them. The expectations and obligations that are established between the teacher and students act to create a culture in the classroom and as such the essence of the setting in which children will learn. These social norms, which are typically established at the beginning of the school year and renegotiated throughout the duration, are seen by the observer to underlie the regularities in the interactions and to form a "hidden grammar" in the classroom (Bauersfeld, 1988). These consistencies, which generally are unnoticed by the participants, nevertheless create different environments for learning.

PATTERNS OF INTERACTION IN MATHEMATICS CLASSROOMS

With this in mind, it has also been contended that in traditional school mathematics classes, the patterns of interaction that are constituted are of such a kind that students do not need to be involved in any mathematical thinking to participate. Instead, they need only to be able to carry out the appropriate behavior in response to the teacher's actions (Bauersfeld, 1988; Voigt, 1985, 1989). As an example, McNeal (1992) found that third-grade students were quite able to participate effectively in discussions on place value to answer adequately the teacher's questions. However, deeper analysis of the discourse revealed that students' comments most frequently consisted of single word linguistic responses which could be given with little if any understanding of place value by simply following the teacher's cues.

It should be realized that questions asked by teachers in school serve a very different function from those questions used in everyday conversation. Teachers usually ask questions for the purpose of finding out if a child knows the answer. However, critical examination of the questions reveals a definite intention to evaluate students' knowledge and accentuate the power imbalance that exists. This imbalance is common in all teaching but is paramount in mathematics. Conversely, in inquiry classes question asking serves a different purpose. From the observer's perspective, it reveals a more equal relationship between teachers and their students. In these settings, teachers ask questions because they truly do not know, they are attempting to get further information, or they are provoking students' reflection. In this regard, these questions act to suggest new aspects of the problem to consider for further exploration by students or to encourage children to re-think their activity. Moreover, teachers may ask a series of questions intended to cause students to reflect on their current understanding and to realize new connections (Ainley, 1988). By definition then, teaching and learning in this sense is viewed as a highly dynamic and reflexive process which necessarily takes into consideration the interactive patterns and processes of communication that are constituted by the teacher and students.

Whole class arrangements such as these in which teaching is viewed as an intensive interactive process and in which talking and reasoning about mathematical ideas are seen as central to learning may create for the teacher a conflict. On the one hand, the teacher would like to create opportunities for an individual student to make personal constructions by asking questions that enable he/she to reflect on his/her own activity. On the other hand, the teacher would like to ensure that those remaining students who are listening are somehow involved in the dialogue. The ways by which teachers reduce this tension are reflected in the nature of the questioning. While the questions are directed to the individual student, they are, at the same time, inclusive of the "collective" student. It is clear that ways of questioning in this regard are quite different from traditional practice; thus their influence on student reflection is realized only when considered in light of the interaction between teacher and students. In this situation, the teachers' role becomes one in which they are asking questions with the intention of encouraging students to reflect on their own thinking. Although it is generally agreed that these questions provide opportunities for more meaningful learning than is available in the traditional initiation-response-evaluation form of interaction, the manner in which teachers relate to their students makes considerable difference in the essential quality of these learning possibilities. With this in mind, let us consider two alternative patterns of interaction and their concomitant discursive exchange.

Funnel Pattern of Interaction

One such pattern that has been described by Bauersfeld (1988) and Voigt (1985) is called the *funnel pattern*. The following episode attempts to illustrate this pattern, the nature of the discourse that occurs, and the meanings that are constructed during the exchange. The lesson under consideration takes place during a class discussion with seven year-olds and is about using thinking strategies to remember basic facts. Conceptually, children who use thinking strategies effectively are thought to have developed abstract or double integration operations (Steffe, von Glasersfeld, Richards, and Cobb, 1983). In the episode described, the teacher has written 9 + 7 = ____ on the chalkboard.

Tch: What does 9 + 7 equal Jim?

Jim: 14.

Tch: Okay. 7 plus 7 equals 14. (She writes 7 + 7 = 14). 8 plus 7 is just adding one more to 14 which makes ____? (voice slightly rising as she writes 8 + 7 = ____ underneath).

Jim: 15.

Tch: And 9 is one more than 8. So 15 plus one more is ____? (voice slightly rising again as she writes 9 + 7).

Jim: 16.

Tch: Good.

This exchange is one that is quite common to those familiar with schooling, and at first blush it appears to be a way to support student learning. In this example, rather than tell the student he is wrong, the teacher does continue the exchange and guides him to the correct answer through a process involving using thinking strategies. The teacher's attempt to keep open the exchange reveals the actual consequences of funneling to the observer. As a participant in the exchange, the teacher initially seemingly accepts Jim's incorrect answer. In actuality she uses it as a starting point to guide him to the correct answer by asking a series of explicit questions which model using thinking strategies. The teacher's intention, in this case, is to orient Jim to the use of thinking strategies as one way to derive basic number combinations. She begins by starting with a known fact (the double, 7 + 7) and increases an addend by one each time (8 + 7; 9 + 7) as a means for solving the problem and arriving at the correct answer.

However, if we begin to examine the nature of the interaction more carefully, we find a situation that is somewhat reminiscent of Wertsch, Minick, and Arns' (1984) account of Brazilian mothers and teachers. In this study, a child was paired with either a mother or a teacher. Each pair was asked to solve a task involving copying a three-dimensional model. While differences did not occur between the adult and child partners in terms of the number of successful completions of the task, clear distinctions did exist in the ways in which they interacted to solve the problem that ultimately influenced the task completion. In the case of the mothers, their only goal was to finish the task correctly and efficiently. In order to succeed, they directed the children to use specific known strategies while they took over the demanding aspects of accomplishing the task. The children were left with only those aspects of the problem which they already could accomplish easily and thus had little opportunity to engage in any of the challenging attributes of the task.

The teachers, in direct contrast, interpreted the activity as a situation for learning. They allowed the children with whom they were paired to choose their own strategies, make mistakes, and correct them. By avoiding specific direction, the teachers created a situation where their children were confronted with problematic situations in which they had to find ways to solve the tasks for themselves rather than rely on their adult partner. Their children were encouraged to engage in the cognitively demanding aspects of the problem and to resolve them for themselves. In attempting to explain the observed differences, Wertsch, Minick, & Arns examined the wider sociocultural aspects and found a relationship between the expectations for children's performance and the production of goods within the social setting. The economic setting of the Brazilian home was such that it was essential to produce products with a limited waste of resources. Hence the attitude and practice in such an environment influenced the mothers to adopt a stance that they could not risk letting children make mistakes. In school, just the opposite occurred as the goal was not to produce an immediate product but to allow students to achieve their own learning through their mistakes.

At first glance, in the funnel pattern, illustrated in the 9 + 7 problem, the student seems to be engaging in mathematical thinking during the exchange, but further analysis of the exchange reveals that the student does not need to understand or even to use a thinking strategy in order to be able to provide correct responses to the teacher's leading questions. More precisely, all the student needs to do to engage successfully in the interaction is to fill in the missing number which will then complete the teacher's sentence. Essentially in each case in the illustrative episode, the teacher's question could be interpreted as, *"What I want you to do is add one to the answer I have just given."* The student can provide a valid answer by simply knowing which number comes next in a counting sequence. It is clear that interacting in this manner does not provide the student with an opportunity to explore the possibility of using thinking strategies or even to engage in finding other ways to solve the problem for himself. In reality, the teacher is the one involved in the mental activity of using thinking strategies as she creates the series of related number sentences in order to generate the final answer. In this interaction, like that of the Brazilian mothers, the teacher is carrying out the demanding aspects of the task to ensure the student's arrival at the correct answer. To clarify, from the observer's perspective the teacher is seen as curtailing the possibility that the student will engage in any meaningful thinking of his own.

Viewed from the observer's perspective, the funnel pattern can be characterized as a narrowing of joint activity to produce a predetermined solution procedure preferred by the teacher. In this case, the nature of the social exchange which consists of providing students with leading questions in an attempt to guide them in their learning, acts to supersede options to create the openness necessary for teachers and students to engage in the process of negotiation of mathematical meaning. Simply put, this leaves students in a situation where they need only to generate superficial procedures rather than meaningful mathematical strategies in order to participate.

Focusing Pattern of Interaction

The *focusing pattern of interaction*, in contrast, is characterized by an exchange in which the teacher's guiding questions act to focus the joint action. Initially, this pattern appears to be similar to the funnel pattern as its intent is to provide opportunities for learning through joint activity. However, the pattern that emerges is quite different as the teacher's intent in questioning is to focus the attention of the student to the critical aspect of a problem, -.to pose a question which serves to turn the discussion back to the student leaving him/her with the responsibility for resolving the situation.

An example of such a pattern is drawn from one of the second-grade mathematics classrooms in which I have been conducting research. In this illustration, the class is involved in a whole-class discussion of their methods for solving two-digit subtraction problems in which the units digit in the minuend is smaller than the units digit in the subtrahend as in the problem 66 - 28. Subtraction problems of this type seem to be particularly difficult for children, and they need to make a shift in perspective in order to find a solution method that makes sense (Wood, 1991). In this episode presented, the class is beginning to discuss the problem 66 - 28 = ____ which is written in a horizontal format. The teacher asks:

Tch: What did you come up with, John?
Jhn: 38.
Tch: 38. (John comes to the front of the class, while the teacher moves to the back).
Jhn: We put 28 under 66. (As he talks, he writes 66 - 28 in the vertical format,) and we took away - we - I took - the 6 and 8 off. (He points to the numerals 6 and 8). And we said there was 60 and 20 there. And if you take away 20 from 60, it's 40. (He puts his finger on the 60 and then on 40). And you still have to take away 8. So we took - there's 46 left over. (He puts his finger on 66). If you take that 6 back and take away that 6. . .(pointing to the 6 in the 46), and that's back to 40. (He holds up his fist). And you still have to take away 2, so 39, (he holds up a finger) then 38 (he writes 38 underneath the equal sign.)

From her past experiences and from listening to students in this particular class, the teacher knows that John's solution is more sophisticated than any of the other students have given earlier in this discussion. She is also aware that to understand and use this solution effectively requires an understanding of tens and ones and a deeper sense of the operation of subtraction. She also senses that John's halting explanation indicates that he may still be somewhat unsure of this solution method.

Self-generated algorithms

As some background to this situation, it should be noted that in our classes the standard algorithmic method for solving these problems are not given by the teacher. Instead children are encouraged to generate their own solution methods. Many students in this class solve addition problems such as 66 + 28 by partitioning the numbers into tens and ones, adding the tens, and then operating with the ones (Cobb & Merkel, 1989). So, 60 + 20 = 80; 6 + 8 = 14; 80 + 10 = 90; 90 + 4 = 94. Quite naturally what follows is that many children anticipate, from their interpretation and understanding of two-digit addition, that the subtraction of two-digit numbers will follow a similar procedure. This notion is initially correct, and students partition numbers and subtract quite successfully until they encounter those problems in which the units digit in the minuend is smaller than the units digit in the subtrahend. Then the situation arises with a problem such as 66 - 28 in which a conflict occurs not when subtracting the tens, but when operating with the ones. Typically, students think of the problem as 60 - 20 = 40 and then encounter the situation of 6 - 8 which is not workable in the previously established way. At this juncture, a very common resolution is that children reframe the problem as 8 - 6 = 2, then 40 + 2 = 42. From their perspective, because they may not fully understand that subtraction, unlike addition, is not commutative, this seems like a reasonable procedure. However, finding this method unsuccessful for arriving at a valid answer, the children are now confronted with a conflict for which they need to find a way to resolve. In the process of trying to do so, they begin to question and rethink their underlying assumptions about subtraction. We can infer from the solution John has given that he has found a way to solve the problem that involves partitioning numbers which suggests that he has also developed some understanding of the properties that hold for subtraction.

The teacher continues the discussion and asks the rest of the class, *"Do you have any questions?"*

Tch: Okay, do it again (to John). He's not sure what you did.

Jhn: We put 66 under the 28. Then we took off the 6 and the 8, and if you take away 20 plus 60 it's 40. And if you put the 6 back on and the 8, we have 46. Then we take away - we still have to take away the 8 then you take away that 6, now you have 40 back and you still have to take away 2.
(All the while that John is explaining, he is pointing with his finger to the numbers, but this time does not touch them.)

Elz: But--but why did you take the 6 and 8 off?

Jhn: It was more easier.

The teacher, looking around the room at the reactions of the children determines that many students still do not understand John's solution. She also thinks that John's response to Elizabeth does not fully answer her question. Perhaps providing a more complete rationale might create an opportunity for him to reflect on his thinking in order to find a way to make his solution understood by others.

Tch: Okay. Could you write down beside it what you did? Maybe that would help us see it. Instead of 66 minus 28, what did you do?

Jhn: 60 take away 20 equals. (He writes 60 - 20 = vertically and looks at the teacher).

Tch: Would you write what you get? (He writes 40 underneath 60 - 20). Okay, what did you do next?

Jhn: Then we put the 6 back on. Then it equaled 46. (He writes 46 next to the 40). And you still have to take away 8; so you have 40 back. And um, if you take 2 away, - you have to take away 2 more; so we got 38.

The teacher realizes that John's initial verbal explanation and pointing to the various numbers as he talked did not seem to help the other students to understand. From the observer's perspective, the request for John to provide symbol notations for his verbal explanation, serves two purposes. First, the request serves to provide John with an opportunity to clarify his own thinking. John is able to extend his verbal explanation and his motor activity of pointing to parts of the numbers to the use of number symbols as a way to provide meaning for others. Second, partitioning the numbers into tens and ones is a common strategy that is already used by many of the students, and notating 60 - 20 is understood. John's use of number symbols on his explanation enables the others to better follow his reasoning. In a subtle way that was not predecided by the teacher, John is becoming aware of the manner in which mathematical symbols can be used to more clearly and concisely express his mathematical activity to others. These events, which arise naturally in settings in which students are encouraged to express their mathematical thinking, create opportunities for students to make connections between their informal explanations to the formal notational methods established by the culture and to begin to realize these symbols can provide mathematical meaning.

Returning to the discussion, the teacher's next question, *"Okay, what did you do next?"* focuses the joint attention of the class to the critical point in his solution and directs John to clarify his reasons for his actions. In this regard, she senses that John's rationale for his activity and use of symbols further illustrates that his reconstitution of 46 has been understood by the class. But, his explanation for subtracting the remaining 8, has not. Further, he seemingly does not have a way to use the number symbols to enhance this important aspect of his explanation.

Tch: (To the class). Make sense? (silence). (Coming to the front of the class). Do you understand what he said about his part? (points to 46 and pauses and looks around at the class). He said, "I have. . ." Let's put 46 up here (she rewrites 46). That's what he has. Then he said, "I've got to go back to 40." (To John). Okay. Why did you go back to 40?

The teacher, summarizing the aspects of his explanation known to be taken-as-shared, then seeks to sharpen the focus of the joint attention of the class. This procedure is meant to create a situation in which the aspect of the problem that is the most critical to understand, and which is not yet understood by most of the class now becomes a topic for discussion. The teacher rephrases his original statement, *"So you have 40 back,"* as if it were a concrete action that John has just performed in an attempt to orient him to the discriminating aspect of his solution strategy for which he must provide a rationale. Expressing his comment as an action he has just performed may enable the others to make sense of his solution a series of concrete actions rather than as an activity with abstract symbols.

Jnh: Cause we took away that 6, - cause you have to take away 8 and you still have to take away 2 more.
Tch: (To the class). You understand how he did that?
Cls: Yeah.
Tch: (Long pause). Very interesting way to do that. Thank you.

In this episode, the teacher's meta comments and questions serve as an attempt to orient the discussion to the aspects of John's solution that are distinctive by not allowing the critical point to fade or change. Most likely, in this situation in which a unique solution has been given, the teacher necessarily anticipates that some students might not understand and asks clarifying questions to keep the collective attention on the discriminating aspects of the solution. At the end of the episode, it is not expected that all students have understood John's method. That they attempted to do so and have raised questions about those points in the explanation that they did not understand is the essence of the interaction.

Summary

Thus far, I have described two patterns of interaction that are distinct from the frequently occurring tripartite pattern found by most studies of classroom discourse. In each of these alternative patterns there are common features that characterize the teachers' role in the discourse. Both the funnel and focus pattern serve the teachers' central intention of trying to create learning situations which enable students to construct mathematical meaning for themselves. The teacher's goal in each case is to aid students in making sense of mathematics by supporting their activity during joint problem solving through continuing an openness not only in the discourse but in the exchange. In the case of the example given of the funnel pattern, the goal for creating a possible learning situation was in guiding students' participation through the process of using a thinking strategy to find the answer to a problem. In the illustration of the focus pattern, the teacher's consideration is on creating situations for learning by maintaining a joint collective focus on the critical and unique features of the problem as determined by the students and on providing an opportunity for students themselves to solve the problem.

The funnel pattern, then, can generally be described as an interaction in which the teacher creates a series of questions that act to continually narrow the students' possibilities until they arrive at the correct answer. In this situation, the teacher recognizes that the student is unable to respond appropriately with the correct answer, and therefore attempts to offer guiding questions for the purpose of enabling the student to solve a problem. The teacher's guiding questions direct the student to the aspects that are important in order to solve the problem (e.g., using the doubles and incrementing by one) and through a series of narrowing queries the student arrives at the answer. This form of exchange always ends with a solution to the problem at hand.

The focus pattern can also be described as a situation in which the essential aspects for solving a problem are brought to the fore. Furthermore, this pattern of interaction can be described as one in which the teacher's inquiries act to indicate to the child the critical features of the problem that are not yet understood. Then the teacher leaves to the child the resolution of the problem. The teacher's role is one of summarizing that part which is commonly thought to be shared and then drawing the students' attention to a critical point not yet understood. This is followed by questioning which first focuses the joint attention and then turns the situation back to the student, letting him/her solve the problem. In this particular interaction, students always have some aspect of the problem still solve.

THEORETICAL PERSPECTIVES ON THE ROLE OF SOCIAL INTERACTION IN LEARNING

The *funnel pattern* in many ways is reminiscent of descriptions of pedagogy provided by those proponents of Vygotsky's theory (Palincsar & Brown, 1984; Lave & Wenger, 1991; Mercer, 1992). One assumption for teaching and learning that is derived from this theoretical orientation is the social influences that foster change are brought about by an apprenticeship model of interaction. In this situation, a novice works closely with an expert in a joint problem solving activity in a zone of proximal development. While gradually accumulating more of execution of the task over time, a child is able to initially participate with a the teacher in skills beyond those she/he is capable of handling independently. Vygotsky theorized that the child internalizes those shared cognitive processes by "appropriating what was carried out in collaboration to extend existing knowledge and skill" (Rogoff, 1990). Furthermore, he believed that qualitative restructuring of thought is related to the acquisition and use of cultural tools and signs for mediating thought. These tools are cultural creations and help to shape the structure and organization of individual thought by emphasizing particular, socially valued, relationships and processes of reasoning. Change in thought, then, resides in the individuals' acquisition and use of their culture through a process of appropriation. These cultural tools and signs are acquired, in part, through socially mediated interventions within the individuals' zone of proximal development as assessed by their success in problem solving situations that were guided by adults or others who have expertise.

Unlike the apprenticeship model, the *focus pattern* of interaction is similar to descriptions of those who are influenced by a Piagetian view of learning for teaching. From this perspective, knowing is an individual, internal process of construction which occurs from the resolution of conflicts or ambiguities between individual existing ways of thinking and experiences with the environment which includes both the physical and cultural setting. Change in intellectual thinking occurs through a process of reflective abstraction involving the mechanisms of assimilation and accommodation. Children learn by modifying their previous cognitive structures as teachers encourage them to clarify their ideas and to consider alternative viewpoints. For proponents of these ideas of teaching, the goal is to create a situation in which different points of view can lead to conflict so that a shift in individual understanding may occur as children consider the alternatives. In this case, it may not only be a conflict between points of view but may be the feeling of discomfort that exists in trying to make sense of ambiguity in a situation that may be created by, say, surprise or miscommunication of meanings.

PATTERNS OF INTERACTION AND SITUATIONS FOR LEARNING

Rogoff (1990), like those mentioned earlier, also notes that different patterns of social interaction occur in accordance with the nature of the situation which in turn creates differences in children's learning. She further argues that the mechanism by which social interaction is thought to influence cognitive development differs according to whether the underlying theoretical orientation and assumptions held are Vygotskian or Piagetian. For Rogoff, differences exist between the fundamental contention of each theorist as to whether development occurs as the result of joint action between adults of expertise guiding children in participation as the starting point or it begins with cooperation between equals adapting their ways of thinking to resolve discrepancies between their views of the world and those held by others. She also suggests that the differences between the two perspectives are related to "differences in the phenomena the two theorists attempt to explain" (p. 140). According to the theory of Vygotsky, children's development is enhanced through social interaction with others who have expertise, usually an adult. In this case, the influence of interaction would be on the development of what Rogoff calls *"understanding and skill."* For Piaget, the nature of the interaction acts to influence a *"shift in perspective"* that occurs as children try to resolve conflicting views and achieve equilibrium. Rogoff continues by suggesting that the variation in the social patterns by which adults and children interact is influenced by the situation for development whether it is a situation of learning that involves understanding and skill or a shift in perspective.

She describes learning, which is understanding and skill, as the "integration and organization of information and component acts into plans for action under relevant circumstances" (p. 142). To this end, she offers learning to tie shoes and to associate elements in order to remember them as examples. She argues that telling children is unlikely to promote cognitive development, but offering mnemonics for remembering a sequence of events "may provide the support, over a number of sessions, to assist the child in learning that skill" (p. 142).

The nature of the interaction that Rogoff presents to enhance development in this regard is one in which the adult plays a strong role in guiding the participation of the child in achieving a skill. In the interaction, described as the funnel pattern, it was the teacher's intention to guide the student's participation in the use of a thinking strategy as a way to associate items to be remembered, in this case 9 + 7. The underlying rationale for encouraging children to use thinking strategies is similar to the one Rogoff describes as *understanding and skill* -- that basic facts can be remembered if one learns to derive an unknown fact through a process of association with a known fact. However, thinking strategies, if understood at a deeper level beyond associating items from memory, enable children to draw relationships between numbers and to recognize patterns that increase their ability to compute and estimate numerical relationships (Rathmell, 1978; Thornton, 1978). In addition, as number combinations become larger, the deeper understanding that students have of thinking strategies enables them to extend their knowledge to find relationships and patterns among 2 and 3-digit numbers and to develop conceptions of place value. An illustration of this is Anna's nonstandard algorithm found in Cobb and Merkel (1989). In the problem 39 + 53 =, Anna made use of the compensation thinking strategy as she explained, "50 plus 30 is 80, then 9 plus one more would be 90, plus 2 more would be 92" (p. 79).

That children may develop a more extensive understanding of thinking strategies requires a situation for learning of the type that Rogoff refers to as a *shift in perspective*. She describes this as "giving up understanding of a phenomenon to take another view which contrasts with the original perspective" (p. 142). Social influences are viewed as enhancing changes of perspective, but require of the learner an understanding of the social processes of *shared communication, intersubjectivity* and *reciprocity*, which Rogoff contends are difficult for children to engage in easily.

Like Rogoff, Kamii (1985) also notes that patterns of interaction that are constituted between adults and children create different opportunities for learning. For Kamii conflict, both cognitive and social, is of central importance for learning. Along with Piaget, she believes that the social relations children construct in order to engage in cooperation are the same as those logical relations they construct to make sense of the physical world. Social influences enable children to learn through the instigation of cognitive conflict and the engagement in logical operations as they attempt to resolve their differing perspectives. In order for this to occur, Piaget considered three conditions to be essential in the exchange: a common scale of intellectual values and means for communicating them, intersubjectivity by which partners recognizing propositions as not self-contradictory look for agreement, and a reciprocity of propositions existing between them. From this perspective, the "socio-affective and intellectual climate of a classroom heavily influences the way children learn or do not learn any academic content" (p. 39). The manner in which conflict is handled in the classroom is seen to be of great importance in the creation of different opportunities for students to learn. According to Kamii, those situations of conflict in which teachers think their role is to direct and organize students will create in children a continuing need to be managed. In their role, students hold the expectation that the teacher will make decisions for them, and thus they feel obligated to follow what others tell them to do.

In situations where conflict or different points of view are encouraged and in which resolution involves giving reasons, students are enabled to develop feelings of autonomy. Teachers who believe their role is to create trust and respect for others' ideas develop in their students the need to be self-directing and self-reliant. Mathematics classrooms in which mutual affection and respect exists among the members and children are encouraged to make decision for themselves create situations for meaningful learning. In contrast, those classrooms in which the climate is such that children are discouraged from resolving conflicts for themselves and expect to be managed create situations in which students will construct less knowledge (Kamii, 1985).

SOCIAL NORMS, PATTERNS OF INTERACTION, AND MOTIVATION TO LEARN

Classrooms in which children interact to present their ideas and points of view necessarily have as underlying social norms those that reflect mutual trust and respect for one another's thinking. If the teacher's intention is to create a climate in which children offer different ideas -- some of which may create conflict -- then children need to be confident that, no matter what, their ideas and decisions will be respected. They need to be assured that they may risk expressing their mathematical thinking to others without fear of ridicule or embarrassment. From the children's point of view, if they are to accept the obligation to make decisions and share their thinking, they expect the teacher to accept and respect their thinking as well. Teachers, therefore, are obligated not to attempt to direct or impose their ideas on their students, but, instead, they are expected to support students in their attempts to consider alternative perspectives and coordinate different points of view. Consequently, if teachers in these classrooms were to advocate to their students ways they might solve problems or guide them through a specific procedure, then they would be in violation of the mutually established classroom norms which could lead students to lose their trust in their teacher (Pimm, 1990).

Conversely, students do expect that teachers will help them in resolving their interpersonal conflicts and support them in their attempts to make sense of mathematics. The way in which the teacher interacts with students to accomplish this reflects the nature of the norms that have been initiated by the teacher and mutually negotiated with the students (Wood & Yackel, 1990). This network of expectations and obligations forms the basis from which individuals develop routines that guide their actions and create the consistency in the ways in which they interact. Patterns of interaction that are constituted are consistent with the norms that are established and create the classroom climate and the ensuing opportunities for children to construct mathematical meaning (Wood, 1991). Thus, differences in learning environments found in schools can largely be attributed to the nature of the expectations and obligations that have been negotiated.

If we want students to attempt to make sense of their mathematical experiences, then as teachers we need to enable students to participate in settings in which they are responsible for communicating their thinking to others and understanding the thoughts of others. Situations of this kind necessarily require the members to engage in interactions with one another that involve the fundamental human social processes of shared communication, intersubjectivity, and reciprocity. These social relations are believed by Piaget to be developed in cooperative social interaction and are parallel to the development of logical relations in children. Thus, he argues that in order for both logical and social relations to develop, children need to participate in situations, including classroom communities, in which the patterns of interactions reflect underlying expectations, obligations which focus on valuing children's meanings and their growth in understanding.

It has previously been contended that classrooms in which the atmosphere is one of mutual affection and trust create situations in which students are encouraged to present their ideas (Cobb, Wood, & Yackel, 1991, 1993; Cobb, Yackel & Wood, 1989). In these settings, it has also been noted that children often create conflict as they present their differing viewpoints. If teachers encourage these conflicts and create opportunities in which students attempt to understand the perspectives of others and to resolve the difference for themselves, then students become responsible for their actions and learning. Children in these communities recognize it is their obligation to provide the explanations and rationale for their point of view. They also realize that they are the ones responsible for explaining their reasoning in a way that can be understood by others. This creates a situation in which children are impelled to think about their actions and what they must do in order to be understood by the others (Kamii, 1985).

The manner in which these possibilities are created is reflected in the way in which teachers interact with their students and the community for learning that is constituted. Teachers who develop ways of interacting such as focusing joint attention on points of ambiguity and then step back to allow students to make clear their thinking create a classroom culture in which children are responsible for their thinking and subsequently are motivated to find ways to be understood by others. It is this essential quality of creating situations in which children are able to find ways to understand and reason about their mathematical activity that distinguishes the social interaction found in those classroom cultures which not only encourage children's cognitive development but influence their motivation for learning mathematics.

NOTES

1. The writing of this paper was supported by the National Science Foundation (#MDR 8850560) and while the author was the Snodgrass Scholar in the School of Education, Purdue University. All opinions expressed are those of the author.

2. Thinking strategies for learning basic facts relies on the notion of using a known combination of numbers to find an unknown fact (e.g., known fact 6 + 6 = 12; derived fact 6 + 5 = 11 because 5 has one less than 6, so the answer must be also one less. These strategies include: increasing or decreasing an addend by one or two; compensation (e.g., add one subtract one); inverse (e.g., 7 + 5; 12 - 5 = 7). Several research studies have shown that using thinking strategies is an effective method for children to develop numerical relationships and to derive unknown facts (Rathmell, 1978; Thornton, 1978; Treffers, 1991).

3. The notion of "focusing" is the result of "wonderfully, enjoyable e-mail discussions" with David Pimm. In addition, some of the ideas central to this paper were elaborated in the course of discussion with Heinrich Bauersfeld, Götz Krummheuer, & Jörg Voigt at IDM, University of Bielefeld.

REFERENCES

Ainley, J.: 1988, Perceptions of teachers' questioning styles In *Proceedings of the Twelfth International Conference on the Psychology of Mathematics Education*, Veszprem, Hungary, 92-99.

Bauersfeld, H., Krummheuer, and G., Voigt, J.: 1988, Interactional theory of learning and teaching mathematics and related microethnical studies In H. G. Steiner and A. Vermandel (Eds.), *Foundations and methodology of the discipline of mathematics education*, Antwerp: Proceedings of the TME Conference, 174-188.

Bauersfeld, H.: 1990, 'Personal Communique.'

Bauersfeld, H.: 1988, Interaction, construction, and knowledge: Alternative perspectives for mathematics education In T. Cooney & D. A. Grouws (Eds.), *Effective Mathematics Teaching*, Reston, VA: National Council of Teachers of Mathematics and Erlbaum Associates, 27-46.

Bauersfeld, H.: 1980, 'Hidden dimensions in the so-called reality of a mathematics classroom,' *Educational Studies in Mathematics*, 11, 23-41.

Christie, F.: 1989, *Language education* Oxford: Oxford University Press.

Cobb, P., and Merkel G.: 1989, Thinking strategies as an example of teaching arithmetic through problem solving, In P. Trafton (Ed.), *1989 Yearbook of the National Council of Teachers of Mathematics.* Reston, VA: National Council of Teachers of Mathematics, 70-81.

Cobb, P., Wood, T. & Yackel, E.: 1991, A constructivist approach to second grade mathematics,' in E. von Glasersfeld (Ed.), *Radical constructivism in mathematics education* (pp. 157-176). Dordrecht: Kluwer.

Cobb, P., Wood, T., and Yackel, E.: 1993, Discourse, mathematical thinking, and classroom practice, in N. Minick, E. Forman, and A. Stone (Eds.), *Contexts for Learning: Social Cultural Dynamics in Children's Development,* Oxford University Press, 91-119.

Cobb, P., Wood, T., Yackel, E., and McNeal, B.: 1992, Characteristics of classroom mathematics traditions: An interactional analysis, *American Educational Research Journal*, 29 (3), 573-604.

Cobb, P., Wood, T., Yackel, E., Nicholls, J., Wheatley, G., Trigatti, B., and Perlwitz, M.: 1991, Assessment of a problem-centered second-grade mathematics project, *Journal for Research in Mathematics Education* 22, 3-29.

Cobb, P., Yackel, E., and Wood, T.: 1989, Young children's emotional acts while doing mathematical problem solving, In D. B. McLeod & V. M. Adams (Eds.), *Affect and Mathematical Problem Solving: A New Perspective*, New York: Springer-Verlag, 117-148.

Edwards, D., and Mercer, N.: 1987, *Common Knowledge* London: Methuen.

Erickson, F.: 1982, Taught cognitive learning in its immediate environments: A neglected topic in the anthropology of education, *Anthropology and Education Quarterly,* 13 (2), 149-180.

Goodlad, J. I.: 1983, *A Place Called School,* New York: McGraw-Hill.

Hoetker, J., and Ahlbrandt, W.: 1969, 'The persistence of recitation,' *American Educational Research Journal,* 6, 145-167.

Holt, J.: 1982, *How Children Fail* New York: Dell Publishing.

Kamii, C.: 1985, *Young Children Reinvent Arithmetic: Implications of Piaget's Theory,* Columbia. NY: Teacher College Press.

Lave, J. and Wenger, E.: 1991, *Situated Learning: Legitimate Peripheral Participation* New York: Cambridge, University Press.

Mehan, H.: 1979, *Learning lessons: Social Organization in the Classroom* Cambridge, MA: Harvard University Press.

Mercer, N.: 1992, Culture, context, and the construction of knowledge in the classroom, In P. Light and G. Butterworth (Eds.), *Context and Cognition*, (28-46) Hillsdale, NJ: Lawrence Erlbaum Associates.

McNeal, B.: 1992, *Mathematical Learning in a Textbook-based Classroom* Paper presented at the International Congress of Mathematics Educators, University of Laval, Quebec, Canada.

Palincsar, A. and Brown, A.: 1984, Reciprocal teaching of comprehension - fostering and comprehension - monitoring activities, *Cognition and Instruction* 1 (2) 117-175.

Pimm, D.: 1990, 'Teaching as meta-communication,' In G. Booker, P. Cobb, and T. de Mendicuti (Eds.), *Proceedings of the Fourteenth International Conference on the Psychology of Mathematics Education*, Oaxtepec, Mexico.

Rathmell, E. C.: 1978, Using thinking strategies to teach the basic facts, In M. Suydam & R. Reys (Eds.), *Developing Computational Skills* (pp. 13-38), Reston, VA: National Council of Teachers of Mathematics.

Rogoff, B.: 1990, *Apprenticeship in thinking: Cognitive Development in Social Context*, New York: Oxford University Press.

Sinclair, J., and Coulthard, R.: 1975, *Towards an Analysis of Discourse: The English used by Teachers and Pupils* London: Oxford University Press.

Steffe, L. P., von Glasersfeld, E., Richards, J., and Cobb, P.: 1983, *Children's counting types: Philosophy, theory, and application* New York: Praeger Scientific.

Stodolsky, S.: 1988, *The Subject Matters: Classroom Activity in Math and Social Studies* Chicago: University of Chicago Press.

Stubbs, M.: 1974, *Organizing Classroom Talk*, Occasional paper 19, Center for Research in the Educational Sciences, University of Edinburgh.

Thornton, C.: 1978, Emphasizing thinking strategies in basic fact instruction, *Journal for Research in Mathematics Education*, 9, 214-227.

Treffers, A.: 1991, Meeting innumeracy at primary school, *Educational Studies in Mathematics*, 22, 333-352.

Voigt, J.: 1985, Patterns and routines in classroom interaction, *Recherches en Didactique des Mathematiques*,' 6, 69-118.

Voigt, J.: 1989, Social functions of routines and consequences for subject matter learning, *International Journal of Educational Research*, 13 (6), 647-656.

Voigt, J.: 1990 The microethnographical investigation of the interactive constitution of mathematical meaning In*Proceedings of the 2nd Bratislava International Symposium on Mathematics Education*, 19 pp. 120-143.

Wertsch, J., Minick, N., and Arns, F.: 1984, The creation of context in joint problem-solving, In B. Rognoff, and J. Lave (Eds.), *Everyday Cognition*, Harvard University Press, Cambridge, MA, 151-171.

Willis, P.: 1977, *Learning to Labor: How Working Class Kids Get Working Class Jobs*, Aldershot, Hampshire: Saxon House.

Wood, T.: 1991, Learning in an inquiry mathematics classroom, in F. Furinghetti (Ed.), *Proceedings of the Fifteenth International Conference on the Psychology of Mathematics Education* (vol. 3, pp. 356-363) Assisi, Italy.

Wood, T., and Sellers, P.: in review, 'Assessment of a problem-centered mathematics program: Third grade,' *Journal for Research in Mathematics Education*.

Wood, T., and Yackel, E.: 1990, The development of collaborative dialogue in small group interactions, in L. P. Steffe and T. Wood (Eds.), *Transforming Children's*

Mathematics Education: International Perspectives, Lawrence Erlbaum Associates, Hillsdale, NJ, 244-252.

Wood, T., Cobb, P. and Yackel, E.: 1991, Change in learning mathematics: Change in teaching mathematics, in H. Marshall (Ed.), *Supporting Student Learning: Roots of Educational Restructuring*, Norwood, NJ: Ablex, 177-205.

Yackel, E., Cobb, P., and Wood, T.: 1991, Small group interactions as a source of learning opportunities in second grade mathematics, *Journal for Research in Mathematics Education* 22 (5), 390-408.

Terry Wood,
Department of Curriculum and Instruction,
School of Education,
Purdue University,
West Lafayett,
IN 47907-1442 U.S.A.

J. EVANS & A. TSATSARONI

LANGUAGE AND "SUBJECTIVITY" IN THE MATHEMATICS CLASSROOM*

INTRODUCTION

The context for much - though not all[1] - learning and problem solving in mathematics is the mathematics classroom at school. To talk about the culture of the classroom is already a move away from "scientific" views which would see the important functions and activities of the classroom as able to be directed according to their criteria of rationality, of understanding, and of what is to count as knowledge. The questioning of these views has had diverse effects: sometimes, a complete rejection of them, unveiling in the process the power relations that such claims to scientificity (e.g, the claim of mathematics to universal applicability) support; at other times, the recognition that the condition of any tradition, including the scientific, is a more or less unified meaning which keeps getting heard anew and reformulated. In this second view, what supports *and* delimits the possibilities of teaching and learning is the *culture* of the classroom. This culture is what makes the activities and the tasks, e.g. problems to solve, *meaningful* - in one way or another.

But what are the important aspects of this classroom culture? It has been argued (Evans 1993, 1994) that what makes these activities or practices meaningful includes the following: their *goals*; the *material and institutional resources* available to the practices; the *social relations*, including the exercise of power, on which they are based; and the *language* or *discourses* which shape the practices themselves; see Sec.1 of this chapter.

We argue that the bases for the culture of the mathematics classroom are the different activities - or we might say, the various discursive practices (in order to emphasise their basis in language) - at play in the classroom. These practices constitute (provide the basis for) the crucial aspects (goals, social relations, material resources) of the classroom culture. These discursive practices are historically located (Foucault 1977; Walkerdine 1984), and socially specific (Walkerdine 1990) in that they position subjects differently. And it is these differences that a teacher or researcher should attend to.

One of the dilemmas of research, as well as of the practices of teaching, is the extent to which the culture of the classroom, the context in which students confront a task is *shared*, or common across all the students who participate, and the extent to which the culture, the context, is *particular* to each student. A number of researchers have discussed the common culture of

S. Lerman (ed.), Cultural Perspectives on the Mathematics Classroom, 169–190.

the classroom, e.g Cobb et al. (1989); see also Wood (this volume). We shall investigate the commonality/particularity problem using a study of adult ("mature") students (Evans 1993).

An approach to describing and differentiating the contexts in which a learner confronts a task, exercise or problem, by pointing to the discursive practice(s) in which the subject has his/her positioning, is discussed with reference to empirical material; see Sec.2.

Two types of analysis of interview material are illustrated. *Cross-subject analysis* shows that, in response to the same question, different subjects called up different practices. Here we examine the results for two interview questions, both requiring the calculation of 10%, one in abstract and one for a 10% tip on a restaurant meal. See Sec. 3.

Case studies, on the other hand, allow the consideration of *particularity* in the student's experience of the classroom. Here we consider in more depth as illustrations the responses of two subjects to the percentage questions; see Sec. 4.

1. SKETCHING IN THE CULTURE OF THE MATHEMATICS CLASSROOM

Our basic approach draws on the work of those US researchers who have acknowledged the work of the "activity theorists" such as Vygotsky (e.g. Cole, Scribner, Lave, Saxe), and that of post-structuralists (e.g. Walkerdine, Taylor). Instead of seeing individual cognition, task, and context as separable, they are seen as a whole. The argument is that the individual (or his/her thinking) cannot be neatly separated from the context, nor can the task be neatly separated from the context (with the latter as "background"). This approach argues that cognition and performance are context specific, in a fundamental sense.

The goals of an activity are emphasised strongly by those following the "activity theorists", as a basic quality of human action; goals also can be seen to include the values of the practice, e.g. the level of precision required in calculations (Maier 1980), and the need for "flexibility" and "efficiency" (Scribner 1984 and Lave 1988).

Some of the authors refer to material and institutional resources as an important part of the activity-in-context. For both Scribner and Maier, the technology of computation is crucial in moulding numerate thinking. A slightly different sense is implied by Lave's description of the importance of the physical layout of supermarkets (as "arenas"), or the physical characteristics of packaged foods, in explaining how subjects make decisions in shopping or dieting practices. Saxe points to the institutional bases of a particular practice, including the artefacts involved in, say, the packaging of candy bars in a locality, and the prevailing patterns of inflation.

Among these writers, there is a universal emphasis on "social relations", though these are understood in several ways: Saxe (1991) emphasises social interaction as a resource in learning; Lave refers to expectations of how subjects will act in particular settings.

Thus, the way that context is understood within activity theory leads towards a recognition of:

(a) the normative quality of mathematics, that is the idea that rationality requires its use in all "numerate" situations - but with it the illegitimacy of its claim to possess the truth for every conceivable practical problem: unlike mathematical/calculative (precise, coherent etc) thinking, practical activities exhibit a different, informal, yet rational, thinking. In these views mathematics becomes a resource for a *practical* subject (with purposes, desires, intentions, etc.) to "freely" draw on.

(b) the idea that practical activities shape values - and therefore what is to be understood by flexibility, efficiency, systematicity, etc - rather than mathematics providing these definitions and their criteria;

(c) a continuing commitment to seeing social and institutional factors as shaping the context of the activity;

(d) a view of the context as "natural" - that is, as able to be described simply by naming it.

Turning to Walkerdine's work, her shift to an emphasis on discourses and discursive practices, on the one hand, is an attempt to show how subjects are regulated in particular practices, thus pointing to differences in power, as well as to "social differences", related to gender and social class. On the other hand Walkerdine brings a systematic, post-structuralist approach to language, in her use of ideas about relations of signification, in order to analyse the ways that meaning is constituted and interpreted.

Post-structuralism, in particular, continues and radicalizes the critique that structuralism initiated of the traditional philosophical view of language. The latter viewed language as a *transparent* medium of communication; that is, language describes an outside reality, or expresses an inner reality of ideas or emotions. Congruent with it is a notion of truth as correspondence. And the traditional view makes the further assumption that private, mental reflection precedes public, intersubjective discourse (Jay, 1982). In contrast, in post-structuralism, language is understood neither as an available *tool* destined to help this or that goal come to expression, nor as a system of *codes*, but as *text* which constitutes itself by drawing everything that touches it - the verbal expressions themselves, emotions, cognitions - into its movement (Hamacher 1989).

2. POSITIONING IN PRACTICES

In our approach, cognition and its "context" are seen as inseparable, and as constituted by the discursive practice(s) in which the subject has his/her "positioning". Therefore, in order to understand cognition it is necessary to specify the practice(s) within which the subject is addressing the problem. Evans (1993) has attempted to produce a synthesis of previous answers to this problem, that would avoid tendencies to overemphasise either the determination of human action - as in some of Foucault's work (e.g.1977) - or alternatively its freely chosen character. Evans' approach involves describing the positioning of a subject confronting a problem as a "resultant" of the practice(s) in which all subjects in that situation *are positioned*, and of the practice(s) which the particular subject *calls up*. His analysis of a set of interviews with adult students (e.g. Evans, 1988, 1991) shows that it is often (though not always) possible to describe a particular subject's positioning in a particular situation - and to understand their thinking and emotions in this context.

In general, the subject's positioning will depend on the interplay of a number of factors, including:

- language and the discursive features of the situation;
- social differences in, say, class or gender terms; and
- the subject's "investments of desire" (Hollway 1989).

In Evans' particular project, the positioning of the students in the interview depended of course on the setting of the research, and on the social relations of the researcher and the subjects within it.

The setting was a U.K. Polytechnic with a relatively high proportion of "mature" students (over 21 years of age, returning to study after some years of work or child-care). Over three years, new entrants to two degree courses were asked to complete a questionnaire including items about previous numerate experiences, and a "performance" scale, followed immediately by a version of the Mathematics Anxiety Rating Scale (Richardson and Suinn 1972). At the end of their first year, a subsample (n=25) were given an interview based on a number of "practical" problems - e.g. reading graphs, deciding how much (if at all) they would tip after a meal out, deciding which bottle of sauce was the "best buy".

Evans' method of interviewing used insights from the ethnographic tradition, (e.g, Hammersley and Atkinson 1983) and from the psychoanalytic critique of traditional field work (Hunt 1989). It attempted to synthesise the approaches of life history and problem solving interviews through the use of *contexting questions* related to each problem. The aim was to study the student's thinking and affect in the interview, as well as in earlier experiences with mathematics and numbers, in relation to her/his positionings in specific practices.

In the interview situation, each subject was positioned as a student on the BA Social Science degree course at the Polytechnic. At the same time, a student's basis for attending the interview itself was that s/he had been "chosen" (in most cases) in a random sampling exercise. Thus the two discourses which provide the overt possibilities for the positioning of the subject in the interview setting were:

- academic mathematics (AM), i.e. college mathematics, or school mathematics (SM) with subject-positions teacher/student; and
- research interviewing (RI), with subject-positions researcher/interviewee.

The letter of invitation, and the script used by Evans as interviewer, were careful to talk about "research" and "interview" and "numbers", rather than "mathematics" or "test". This was an attempt to shift the discourse and the positioning from one having to do with academic maths, to that of a research interview. What Evans called the research interview discourse can be considered to offer a space for other discourses to be called up, in a way that the academic mathematics discourses might not do. To the extent that a practice is called up other than AM, the subject will have access to ideas, methods of reasoning, "skills", and emotions from that practice. One aim of the interview, then, was, as much as possible, to create a situation with space for the subject to call up one or more discourses - just those that would be called up if the subject were to be confronted with the problem in the course of his/her "everyday life".

Unlike most similar interviews conducted in mathematics education or psychological research, each problem was introduced by the *current* (C) contexting question, which asked specifically "which of your current activities it [i.e. the problem] reminds you of". And the discussion on the problem was rounded off by the other, the *recall* (R) contexting question: "which of your earlier experiences it reminds you of". These two questions, especially R, often brought up much life history material, and would tend to position the subject in the research interview (RI) discourse [2].

For judging the subject's positioning, Walkerdine has shown how to use indicators for determining what sort of discourse is in play, for a particular task in a particular context (1988, pp.53ff). She suggests attention to:

(a) the explicit form of the tasks, in their "discursive features", viz. the terms and constructions used; in Evans' research, this would require examining how the interview itself was described (see above), and how each task or problem was introduced, in the interview script;

(b) the unscripted aspects of interaction between researcher and subjects: e.g. the (conscious or unconscious) emission of different verbal or vocal signs for "correct" and "incorrect" answers; and

(c) comments and responses given by the student both during the problem-solving phase of the interview, and afterwards: e.g. the responses to the contexting questions.

In addition, we would add the following to the list of sources of indicators:

(d) reflexive accounts: on the "history" of Evans's social relations with the students in each year as a group, and with individual students, e.g. the ways in which he had been in the position of "mathematics teacher", or "researcher" or otherwise related, to each of the subjects; and

(e) the setting of interview - its location in Evans' office at college, the arrangement of furniture, the use of a tape-recorder and where it was placed, etc.

Sometimes it was relatively straightforward to make a judgment as to the subject's predominant positioning at a particular point in the interview. Sometimes it was not at all easy to decide. In addition, the subject will generally have a multiple - or interdiscursive - positioning in some "mix" of discourses - in this case, of the academic mathematics and the research interviewing discourses. And the resultant positioning may be fluid and changing. Hence the task is to describe this positioning during crucial episodes of the interview.

In this paper, we focus on: (a) cross-subject analyses of the interviews and (b) within-subject or case study analyses of the interviews, but not on between-subject analysis of the questionnaires (see Evans 1993; Evans, 1994). The cross-subject analyses aimed to produce summaries of results considered comparable across the sample of 25 subjects. The case study analysis of interviews included description of illustrative episodes, plus in-depth analysis of the flow of the interview, and of the webs of meaning linking practices, cognition and affect.

3. ANALYSIS OF "SHARED CULTURE": CROSS-SUBJECT ANALYSES FOR PERCENTAGE PROBLEMS IN THE INTERVIEW

All these students were doing a mathematics course as part of their first year studies in social science at a Polytechnic in the UK in the mid-1980s. We can check whether the culture of the classroom would have been "shared" in a very basic way, by asking whether all students called up the same "predominant" practice when addressing any of the interview questions. Let us consider questions 2 and 4, both "10% of" questions. For the full details of each question, see the Appendix.

For the results of the practices called up for both questions, cross-tabulated, see Table 1. Evans argued that in general a particular subject, confronting a particular problem in the interview, was positioned interdiscursively - in some "mix" of academic mathematics, research interviewing and perhaps some other everyday practice - rather than in a single practice. However, for the cross-subject analysis he aimed to record the practice which *predominated* in each subject's positioning, if possible.

Indicators for a predominant SM positioning were considered to be:
for either question, the use of written calculations; *and/or*
for Qu. 2, expressed confusion as to where to put the decimal point; *or*
for Qu.4, the giving of an answer which involved a fraction of 1p (other than 1/2p). [3]

On the other hand, some subjects were considered to have a predominant positioning in what Evans called practical maths (PM) discourses, i.e. everyday practices with numerate features, such as spending money or eating out. Indicators for a predominantly PM positioning were:
- for either question, the use of mental calculation; *and/or*
- the formulation of an answer in terms of practical (e.g. money) units - especially for Qu. 2, which had been posed in abstract terms.

Table 1 Positionings for Interview Qu.2 [10% of 6.65] and for Interview Qu.4 [10% tip on meal "chosen" from menu]: Cross-tabulation of Numbers of Subjects

		Positioning for Qu. 4 in interview		
		School Mathematics (SM)	Tipping Practices (PM)	Total
Positioning for Qu. 2 in interview	SM	5	12	17
	PM	0	6	6
	Total	5	18	23

Notes:
1. This table excludes 2 of the 25 subjects for whom either response is not available.
2. For the indicators of SM and PM positioning, see the text.

When we consider the results for Qu. 2, the "abstract 10%" question, we see that the majority, but not all (i.e. 17 of 23), of the students, have a positioning predominantly in school maths (SM) discourses, with a minority "in" practical maths (PM) discourses. For Qu. 4 (see App.), the 10% tip, the majority called up out-of-school discourses ("PM"), notably "eating out" - but there were still at least 5 subjects whose positioning was, at least "predominantly", in academic (school/college) maths. For example, interviewee no. 21, a former manager who was unenthusiastic about the menu shown (since she was vegetarian) gave an answer that would have been "inappropriate" in the eating out discourse - namely "25.3p" (for a 10% tip on a meal costing £2-53); but she "corrected" it to "26p" when prompted. Thus we can see that, both for Qu.2 and for Qu.4, the students in the sample address the problem in varying ways, since they call up different practices[4].

Turning to a second cross-subject analysis, for Qu.4, the gender differences in correct performance - 10 of 11 correct for men and 9 of 12 for women - were very small; see Table 2. Nevertheless, differences in positioning may help to explain these, and they are interesting in themselves: all 11 men were classed as calling up eating out for this part of the question, whereas 5 of the 12 women were classed as positioned in academic maths. And only 3 of 5 calling up SM got the question correct as compared with 16 of 18 calling up eating out. Thus, the gender differences in performance for Qu.4 (though very small, and therefore needing replication) appear to be explainable in terms of the predominant positionings of the subjects.

Table 2 Performance on Interview Qu. 4 [10% tip on meal "chosen" from menu]: Cross-tabulation of Number of Questions Judged Correct by Gender and Predominant Positioning

Predominant Positioning	Men	Women	Total
Practical Mathematics (eating out)	10 of 11 (91%)	6 of 7 (84%)	16 of 18 (89%)
Academic Mathematics	-	3 of 5 (60%)	3 of 5 (60%)
Total	10 of 11 (91%)	9 of 12 (75%)	19 of 23 (83%)

Notes:

1. This table excludes two of the 25 subjects who were not asked to attempt Qu.4.
2. For the indicators of SM and PM positioning, see the text.

4. PARTICULARITY IN THE CLASSROOM: TWO CASE STUDIES

In the case study analysis we can be more precise about a subject's positioning and the fact that it is normally interdiscursive (see above). This realisation can affect how we interpret both the thinking and the emotions, expressed and "exhibited" (Evans 1993), of the subjects.

4.1 Jean

Jean was aged 18 at entry and working class[5] with CSE's[6] in both Mathematics and Arithmetic. She had a part time job in a pub at the time of the interview.

She seems to call up school mathematics or college mathematics for Qu. 2 [10% of 6.65], and the everyday practice of eating out, as well as school (or college) mathematics, for Qu. 4. [7] She always seems to get percentages "the wrong way round". In one of her attempts at Qu. 2, she tries {10/6.65 x 100}, but then realises that is incorrect, and gives the answer "0.65", which she says is "just a guess". [8] How did she come up with that? "I just moved the dot".

What may be a conceptual "problem with percentages" shows up most clearly in the second part of Qu. 4. After she has illustrated her tipping practices by saying that for a meal costing £2.75 she would leave £3, she attempts to respond to the question about a "10% tip on £3.75". First, she tries {10/3.75 x 100}, as for Qu. 2, and then, she tries {3.75/10 x 100}. But she doesn't really know: "it goes something like that". The discussion of Qu. 4 ranged over tipping in general, not having sufficient money, having to tip 15% in the USA (for which she was leaving the following Sunday), having to re-learn how to calculate percentages, especially 15%...

In the interview as a whole, the main affective themes are anxiety, diffidence, and worry, constantly expressed or exhibited. She begins diffidently: "I sound horrible on tape"; this may be *exhibiting anxiety* about the *interview*. Then concerning her CSE Grade 3 in Mathematics: "I wasn't very good at all". She also *expresses* a great deal of anxiety about *percentages*; e.g."I always get the formula wrong...", etc., etc.; "I'm going to have to learn percentages again..." (for the USA); "I always mix it up". These can be read as examples of "self-defeating self-talk" (Tobias 1978).

But it is striking that she seems to express even more anxiety about *money*. For example, "I've never, ever got enough money" (for tips); about the level of tipping required in the U.S.A to which she is about to depart, and about being able to afford 15%. She is also anxious about being able to afford the trip at all: "I haven't had a holiday for three years, so this will get us to America: it's the only way I can ever make it...". When shopping, "I do always follow the prices,... for fear of being ripped off", and with wage slips, "Yeah, I follow them through as well". What might have appeared at first to be "mathematics anxiety" seems now to be part of the fabric of her constant worry about money and financial constraint.

We can interpret her responses to Qus. 2 and especially 4, the percentage questions, as errors. And we can read her errors, particularly in Qu. 4 about tipping, as related to a conceptual problem about percentages; and/or a complex of factors (cf. Ginsburg and Asmussen 1988) which may include:
- beliefs about herself as a solver of mathematics problems;
- mathematics anxiety, especially about percentages;
- some anxiety about the interview; and
- anxiety about the relevant practice, viz. tipping in restaurants, and/or tipping in the USA. The latter anxiety also relates to an apparently chronic anxiety about money. Here social class might be conceived as a further explanatory factor.

4.2 A second reading of Jean's interview

Jean's account can be analysed, as in Sec. 4.1, in a rather straightforward way: she makes "errors", but apart from purely conceptual misunderstandings there are a number of *factors* to consider which would help the researcher modify his/her first judgment that it is simply a question of a mathematical error. The research situation is here formulated in terms of:
(a) clear criteria of what is a correct/incorrect answer;
(b) a distinction among errors resulting from an mistake or a slip, errors resulting from a misapplication of an algorithm, and errors based on a lack of conceptual understanding;
(c) the subject is surrounded by a context, and this context is influenced by a

number of factors;
(d) some of these factors are affective as opposed to (or separate from) cognitive; and
(e) the student is supposed to hold beliefs about mathematics and about herself as a mathematical problem solver, etc., which, in turn, influence her performance.

However as is usual in doing empirical research, more subtle and complex reformulations open up different dimensions of the research problem.

To begin with, some of Jean's beliefs about mathematics can be captured as a phenomenon which might be called "classification" after Bernstein (1971) - though there are important differences. Bernstein uses the concept of classification to describe the strength of boundaries between curricular subject contents in educational knowledge. These subject contents are *socially* recognised (implicitly or explicitly), and have effects on the specific identities of teachers and students; i.e., they determine that the subjects accept their authority and values without resistance. This can be contrasted with the boundaries involved in the sort of "distinguishing"/pigeon-holing we are pointing to, which appear to refer to the particular subject, and - to some extent at least - to *flow from* his/her identity, based on formative experiences and "investments"; these boundaries thus are both profoundly affective and cognitive.

Bernstein also suggests that any

> collection code [i.e. system with strong classification] involves an hierarchical organisation of knowledge, such that the ultimate mystery of the subject is revealed very late in the educational life. [...] only the few *experience* in their bones the notion that knowledge is permeable, that its orderings are provisional, that the dialectic of knowledge is closure and openness. For the many, socialization into knowledge is socialization into order, the existing order, into the experience that the world's educational knowledge is impermeable. Do we have here another version of alienation? (Bernstein, 1971, p.57; his emphasis)

In mathematics, the hierarchical organisation of knowledge seems especially marked, and hence the impermeability of the subject, or at least parts of it, may provide the basis for seeing it, or parts of it, as *alien*. But the *affective charge* of that alienation is provided by the formative experiences and the investments that are part of the history/biography of the particular subject her/himself.

Classification and framing, as the structuralist reformulation of (e) above (the subject's beliefs about mathematics), bring to the fore the complex social and institutional processes constitutive of knowledge representations - inside and outside schools - and consequently the socially structured nature of what appears to be a student's likes/dislikes of mathematics. Yet, with structuralist approaches, it is the "deep structure" of a social system, i.e., its class character, expressed in the linguistic codes, which forms the specific identities of the subjects.

However, approximately half of the 25 students interviewed made a distinction between (put simply) "good mathematics" and "bad mathematics". For example, Jean, having done both CSE Mathematics (Grade 3) and CSE Arithmetic (Grade 2), distinguishes the latter as "useful", "should be compulsory" and the former as "not useful", "I don't like it" and "should be an option". She also distinguishes the topics in 1st year mathematics at the Polytechnic in the same way; e.g. percentages, graphs and statistics are relevant, but gradients and algebra are not. This distinction is also expressed in slightly *different* forms by several other subjects.

The above points to the fact that it is indeed possible to analyse Jean's account in a second straightforward way: rather than a number of factors, we might focus on a *determining* factor, namely social class. However, there is something else in her discourse which might attract the attention of the researcher. Listen to her talk: "it goes *something like* that"; "I sound *horrible* on tape"; "I wasn't very *good* at all"; "I always get the formula *wrong*"; "I always *mix it up*". We hear "uncertainty", "anxiety", but all we have are *indices* of something other than language. We might quickly disregard them, calling them "self-defeating self-talk" (Tobias 1978). Or we might pay attention to these signifiers marking her talk. This could lead us to discourses such as the discourse of norms (right/wrong), of aesthetics (good/bad), of science (truth/falsity), all "mixed up" in her talk. Meanwhile, her talk is asserting the usefulness of mathematics as *practical*. It is the following up of this thread of signifiers that blocks any attempt to simply apply the criteria of true/false which normally direct our judgment with reference to the application of a mathematical formula. Instead, we, as researchers, might have to aattend to the discourses providing the basis for her talk.

We can continue listening to her narrative about her trip[10] to America, and her anxiety about it. This would initiate a shift from *a language of conceptual divisions* to a *psychoanalytic language of "desire"*. (We can see this even more readily in Ellen's case, below).

To summarise, one can analyse Jean's case in terms of factors and see social class as the determining factor. But what we are pointing to here is the play of language; and the sensitive attentiveness of the teacher/researcher to this language makes it a thread of indices. It becomes a question of how to *read*. For the researcher has a sense that there is something there to interpret. And while we are attempting to understand, our attending to her language moves us from a language of Kantian concepts and divisions (theoretical-scientific, practical-ethical, and aesthetic) to a psychoanalytic language of desire (e.g. Thom 1981; Ulmer 1986; Laplanche and Pontalis 1973, pp. 481-484).

How is the research process to take cognizance of such a move? Let us consider Ellen's case study.

4.3 Ellen

Ellen was aged 19 at entry, middle class (by parent's occupation), with A-level in mathematics[6]. A student of Town Planning, she worked part-time, currently as an electronics assembler, and previously in a shop (Evans, 1991).

In the interview Ellen expresses overwhelming confidence after almost every question, e.g. for Qu. 1 [reading a pie-chart]: "very familiar, know exactly what it means... don't have to think about it..." [8] - except for Qu. 4!

For most questions Ellen seems to have a predominant positioning in school mathematics (SM), e.g. for Qu.2 [10% of 6.65]; this was done in her head, and correctly. Qu.4, where she was first asked to "choose from the menu", seems to call up the practice of eating out at restaurants. However, she reverts to using pencil and paper - indicative of SM - to calculate a 15% tip. That more than one discourse may have been called up for Qu. 4 is an illustration of being "positioned inter-discursively" (e.g. Walkerdine) or of the "proportional articulation of structuring resources" (Lave 1988). Both Qu.4 [15% tip on a meal of £3.53] and Qu.5 [9% increase on wages of £66.56] were done with pencil and paper, ultimately correctly, though Qu. 4 begins with a slip -see below.

Qu. 4 deserves more detailed consideration, particularly of her answers to the contexting questions (see Sec.2). When Evans asks if she ever goes to a restaurant with a menu like that shown, she seems to reply very quietly and hesitantly. After she "chooses" the seafood platter [£3.53], he asks how much she would tip for a restaurant meal: she replies, somewhat hesitantly, "...well, 15%, I suppose...". [10]
When Evans asks what a 15% service charge would be, she says "Well, I'd have to use pencil and paper". Then -

S : [7 sec. / inaudible / coughs / 6 sec.] Well, 23 1/2 pou- no, that's wrong... [12 sec.]...what I've done wrong, oh (JE: Is it wrong?) Yeah, umm [laughs nervously]... I don't know what I'm doing...
[She realises she has *divided* 15% into £3.53, instead of multiplying]
(JE [2 lines])
S: [15 sec.]... 52.95, 53 pence.

She explains that she rejected the answer produced by dividing because "I just saw that it was obviously not right... it was far too small".

Further on, in response to contexting question (R), she indicates that in restaurants, "I don't usually pay". But she "looks at prices and things,... add them up in my head...". Even if she's not paying?: "I don't want to be an expense".

Her slip in Qu. 4, and what she says in response to contexting question (R), are suggestive of anxiety. Possible interpretations of this include:

(a) anxiety to do with the interview itelf, which she is experiencing as an evaluative situation;
(b) anxiety experienced about the question itelf, since she is not confident about doing a slightly more complex calculation for Qu. 4 (a 15% tip) than she has so far had to do[11];
(c) anxiety about doing the right thing in a restaurant - which comes up before the 15% calculation, and interferes with her doing it.

Support for (c) comes from her hesitation, etc. at the presentation of the restaurant menu; and "I don't want to be an expense" (see below). Counter-indications for (a) and (b) are provided by the next question: she had to calculate a *9%* pay increase, also on paper, also a "complex" calculation, but she got it right.

The anxiety evinced might appear at first to be "mathematics anxiety", since it appears while doing something that an observer *could* choose to label as "mathematical" - though it could also be labelled as "being interviewed" or even as "talking about eating out", etc. However, on reflection, calling the anxiety "mathematical" would be accurate only if we were to assume that her positioning was solely in a school mathematics discourse, as for (b). It would not be accurate if, as we argue here, her positioning is inter-discursive - that is, in "eating out", as well as in SM, and as interviewee.

Her performance appears very competent in the interview (though many of the questions should not be difficult for someone with A-level mathematics) - except for the slip on Qu. 4. In terms of affect, she gives a picture of overwhelming confidence about mathematics and the use of numbers, but seems to show anxiety when she applies the incorrect operation for tipping. Is it possible to interpret her expression of confidence as a defence against anxiety?

4.4. A second reading of Ellen's interview

This last point suggests a need for insights from psychoanalysis in analysing the interviews, here done within an ethnographic perspective. Hunt (1989) suggests some guiding principles:

> (a) much thought and activity takes place outside of conscious awareness; thus, everyday life is mediated by unconscious images, fantasies, and thoughts - which sometimes appear as jokes, slips, dreams, or subtly disguised as rational instrumental action;
> (b) the unconscious meanings which mediate everyday life are linked to complex webs of signification which are ultimately traceable to childhood experiences; and
> (c) intrapsychic conflict (among desires, reason, ideals, norms) is routinely mobilised vis-a-vis external events, especially if they arouse anxiety or link to unresolved issues from childhood. (p.25)

A very general implication of (a) and (c) is that any product of mental activity - including interview talk - may, upon deeper investigation, reveal hidden aggression, (we would add) suppressed anxiety, forbidden desire, and defences against these wishes (p.25). An important implication of (b) is that transference, the imposition of "archaic" (i.e. childhood) images onto everyday objects (people and situations), is a routine feature of most relationships, including fieldwork relationships.

Now we take a second look at our initial reading. First, her "confidence". Her repeated expressions of confidence may cover up some amount of anxiety - exhibited e.g. by the instances of nervous laughter (see e.g. the interview transcript quoted above) - and these expressions can be seen as a defence against anxiety (Laplanche & Pontalis 1973, pp.103-111).

Second, her "slip" in question 4. Hunt's assumption that unconscious images and thoughts sometimes appear in jokes, "slips", etc. supports a conjecture that anxiety was triggered by the presentation of the menu - rather than by being asked to calculate a 15% tip later in the same episode. And that it was anxiety about the context of eating out, perhaps about the relationship(s) in that context. We might next conjecture that the fact that she made a slip is related to these anxieties. And the *content* of the slip might be related to these anxieties, too: the latter, involving division rather than multiplying, led to a result that was *smaller* than it should have been. When we remember that she later admits to wanting not to "be an expense", we might say that her slip was "motivated" by the anxiety. Thus we can note the role that *unconscious* anxiety may possibly have played in the sequence of responses made by this subject.

Third, her anxiety about being a "expense". Here we can illustrate the links between the linguistic ideas of *metonymy* and *metaphor*, and the phenomena of *displacement* and *condensation* respectively; the importance of language/discourse in psychoanalysis was emphasised in Lacan's work (e.g. Thom 1981)[12]. The idea of being an "expense" may be linked - metaphorically - in this woman's history, with that of being a burden in a relationship, one that is infused with desire (Henriques et.al. 1984, Sec. 3). Because of anxiety, guilt, pain, etc. associated with this, the signifier is likely to be "suppressed" (Walkerdine 1988, Chs. 9 & 10). When a problem is presented involving choosing from a menu - with prices attached! - and when she is asked to calculate the amount of a tip, this calculation will be linked - metonymically, through the idea of summing - with the signifier "expense". This key signifier is thus located at the intersection of two, at least, discourses, and there is a play of meaning across its different senses. We could say that multiple meanings are *condensed* on the signifier 'expense". Also, the linkage between the two discourses allows the strong feelings based on desire in the discourse(s) around her relationship and "eating out", to be associated with this particular problem - which appears so very mathematical! Yet we can say that this subject *displaces* her anxiety about being an expense by moving along the chain of signifiers - from "burden" to "expense" to the calculation - and focuses on the sum!

Finally, Evans' departing from his normal script for Qu. 4 to give her a more difficult question than other subjects. Here is where the *reflexive* account for Ellen can be useful.

> I was not aware of having met this student before the interview; our contact would have been confined to the 1st year Maths lectures that I gave. In the interview, I was struck by the fact that she had A-level maths (attained by only 7% of that year's entry) and was convinced at first by her expressions of confidence. I was concerned about how the interview was going, and especially that she might be bored with such easy questions. Then came her responses to Qu. 4. Here, given the "opening" by her mentioning 15% as a tip, and with the above "reasons" in play, I asked her what a *15%* tip would be, likely a more difficult problem than that [10%] asked to other subjects. (Evans, 1993, pp.340-41) [10]

With psychoanalytic assumptions in mind, we can see that the standard "reflexive account" needs to be augmented. On reflection, Evans was able to recall feeling some anxiety himself: he had not met Ellen before, unlike most of the other interviewees. The decision to give her a more difficult problem can be seen as "motivated" by *inter alia* an anxiety that she would find the problems too boring and that the interview would not be a "success". This can be seen as an example of *transference*, the subconscious reaction of the researcher to the subject, based on the imposition of images onto her - e.g. boredom - for which there was no substantial evidence.

By bringing in a different theoretical frame we show the complex character of Ellen's case. This is achieved by considering:

(a) feelings which are not separate from or outside of language/discursive practices which position her as woman, as student, as "poor", as interviewee, etc;

(b) context not as a natural setting "out there" (classroom, restaurant), but as constituted by practices, which are also constitutive of the "individual's" subjectivity; and

(c) language, not as simply representing preconstituted states of affairs but as actively producing them.

That is to say, shifting attention from consciousness (intentions, etc.) to semiotics, from a pre-constituted subject representing a world of (ideal) objects (e.g. mathematics) to signs, there has been an opening up in the problematic of research. In this "second reading" following post-structuralist ideas such as the idea of meaning as a play of signifiers, and psychoanalytic concepts such as condensation and displacement, a much richer explanation is constructed.

Furthermore, this shift was also important in that it led to:

(d) an enhanced awareness regarding the assumptions about mathematics (and its relation to language) which are made by mathematics teachers and researchers. That is to say, the realization that the turn to language might also put in question a fundamental cognitivist assumption of research, namely that mathematics is unequivocally a *closed* conceptual *system* of representation. (see Lerman, this volume, and Tsatsaroni and Evans, 1994).

In addition, this shift in attention also confirms our awareness of the importance of the frames of analysis that researchers (and teachers) bring to research. In this particular case, the decision to draw on the Lacanian psychoanalytic discourse means that the theoretical scheme of analysis relies on and accepts a re-reading of Freudian psychoanalysis through linguistics, and a re-reading of Saussurean linguistics through psychoanalysis. This means that all substitutions and metaphorical links are read in terms of, and therefore reduced to, the psychoanalytical language of desire. In Lacan's account desire arises from a lack, the lack of the phallus, which is the transcendental signifier uncovered by psychoanalytic inquiry. In this tacit way Lacan's work restores "subjectivity" as a "metaphysical" assumption; see,Nancy & Lacoue-Labarthe's (1992) deconstructive reading of Lacan.

CONCLUSIONS

Our approach has, we think, important implications for discussions on three levels:

- on assumptions built into research on mathematics education;
- on "what is mathematics", and what counts as mathematical knowledge; and
- on pedagogy.

Regarding the first, we hope we have conveyed a sense of movement in the research process. It can be said that every new reading of research material has meant the deconstruction - in a loose sense - of certain, taken for granted approaches and assumptions in educational research. Of course, complicated frames of analysis which open up research situations to different readings, as are used in this chapter, do not exclude the possibility that answers to mathematics questions are simply errors, or are due to gaps in knowledge. Rather, it is the unthinking attribution of error we wish to problematize here. A good teacher or researcher must entertain different possibilities; and it is only judgment - and what can support this judgement - that is at issue.

Our approach has been based on a poststructuralist view of language. This view of language has important implications for research and for teaching and learning. We can base our analysis neither on, say, codes or the context of use: the context itself - its essential importance notwithstanding - cannot be *delineated absolutely*. The most trivial examples, whereby attempts are made to make mathematics "meaningful" (cf.Adda 1986) indicate that every time a teacher reaches outside of mathematics to furnish examples which will illustrate the case, the mathematics is "at risk". That is, it is open to the fundamental character of language and of the signifier: the signifier can be repeated and, in so doing, it can break with every given context "engendering" and being inscribed in new contexts without limit (Derrida 1988, pp.12 and 79).

On the second level, we can simply note that the fundamental character of the signifier provides an essential condition which prevents mathematics from being a self-contained, closed system: rather it is a concatenation of signifiers with connecting lines reaching out in all directions; and those links cannot be cut off or excluded, either by an *authorial intention* e.g. of the teacher, nor by *a priori* rules of mathematics as a *system of concepts*. This relates to Walkerdine's (1988) argument that the "mastery of (mathematical) reason" would have to be premised on the supression of extraneous meanings - but that this mastery is based in fantasy. The question of "what mathematics is" is considered in more detail in Tsatsaroni and Evans (1994).

On the third level, we feel that our work implies much for the practices of teaching in the classroom, as well as for research. For example:
(i) The technique exhibited here of ascribing the student's positioning through e.g. the language s/he is using can be useful in diagnosing the reasons for a student's "misconstrual" of a problem to be solved.
(ii) The teacher needs to be *aware* of multiple meanings and (possibly suppressed) affective charges related to ideas drawn on in "making mathematics meaningful".
(iii) All this suggests that the teachers need time to know their students, and that classes cannot be too big. Also, we need to be wary of the "technologization" of teaching and assessment of learning, in particular of the excessive use of pencil-and-paper tests, as in current versions of the National Curriculum in the U K.

NOTES

* We would like to acknowledge the crucial influences over the years of Valerie Walkerdine and John Hayes in our being able to formulate the problems discussed here in these terms. Of course we are responsible for the formulations in this article.

1. For a discussion of "informal" or out-of-school learning in maths, see e.g. D'Ambrosio (1985), Carraher (1991) and Schoenfeld (1991).

2. However, the numerate problem itself also included what might be called a "pseudo-question" (Walkerdine 1988), as used in pedagogic or testing discourses, in that Evans "knew the answer" (or at least thought he did, at the start of the interviewing). This might tend to position the subject in the academic mathematics discourse, rather than in the research interview discourse. This dilemma, this possibility of different positionings, needed to be assessed in relation to particular problems for particular subjects; see Sec. 4.

3. The 1/2p was just being phased out in 1985 and 1986, when the interviews were done.

4. Note further that the bare majority - 12 of 23 -called up a *different* practice for the two problems. This provides contrary evidence to traditional views that, because both Qu.2 and Qu.4 "are" (essentially) "10% of" questions, they both will be (or should be) thought about using the same framework.

5. There were no social class items in the questionnaire for Jean's year of entry but Evans judged her to be working class on the basis of her regional accent, and an often-voiced concern about never having enough money.
6. The CSE was introduced as an alternative qualification at age 16 to the more "academic" GCE O-level in the 1960s in the U.K. The two were replaced by the single GCSE 16-plus qualification in the late 1980s. A-level remains a very specialised exam at age 18-plus.
7. A basis for understanding her calling up of school mathematics and mathematics tests early in the interview may be provided, at least partly, by the unscripted aspects of Evans' performance as interviewer: e.g. offering her paper for Qus.1 and 2 - suggesting that she might want to use written methods, considered here to be indicative of school mathematics-based thinking.
8. All quotations in Jean's and Ellen's accounts are from the interview transcripts; see Evans (1993).
9. The student's reference to her trip to America might perhaps reflect back and be reverberated in the researcher's anxiety regarding his/her topic of concern. For example, mathematics as a conceptual system makes a detour - goes outside itself - by drawing on examples, by relying on language to convey its basic concepts, to be understood. About the risks inherent in such a move, see Tsatsaroni and Evans (1994), and also Derrida (1978) and Caputo (1987).
10. She was one of the very few students to suggest a 15% tip as normal.
11. Almost all students were asked to calculate a *10%* tip at this point in the interview, though of course she would not know this.
12. For further on Lacanian psychoanalysis see Lacan (1977) and Ulmer (1986).

APPENDIX

Details of the 10% questions used in the interview

Qu. 2: "abstract" 10%
(C) Does this remind you of any of your current activities?
What is 10% of 6.65?
(R) Does this remind you of any earlier experiences?

Qu. 4: 10% tip on selected restaurant meal
(The problem was introduced by reading out several contexting questions - CA, CB and A)
[Show the "menu" (see Fig. 1)]
(CA) Do you ever go to a restaurant with a menu anything like this?...
(CB) Would you please choose a dish from this menu?...
(A) Suppose the amount of "service" that you leave is up to the customer: what would you do? ...
(B) Could you tell me what a 10% service charge would be?...
(R) Does this remind you of any earlier experiences?

The "menu" for Qu.4 in the interview.

CHICKEN Served with sweet corn, banana fritter, bacon, fresh tomato,
MARYLAND whole French beans, jacket baked potatoes with sour cream and chives or French fried potatoes.

Roll and butter.

Ice cream, or a selection from our cheese board, biscuits and butter.

£3.75

SEA FOOD PLATTER Served with tartare sauce, whole French beans, jacket baked potatoes with sour cream and chives or French fried potatoes.

Roll and butter.

Ice cream, or a selection from our cheese board, biscuits and butter.

£3.53

GRILLED TROUT 10 OZ	Served with tartare sauce, whole French beans, Jacket baked potatoes with sour cream and chives or French fried potatoes. Roll and butter. Ice cream, or a selection from our cheese board, biscuits and butter.	**£3.81**
Coffee	Special blend black or with cream	**27p**
Connoisseur Coffees	Served in large goblet glass with cream: Irish (*Irish Whiskey*), Caribbean (*Rum*), Russian (*Vodka*), Parisienne (*Brandy*), Calypso (*Tia maria*) Highland (*Scotch Whisky*) Mine Hosts (*Cointreau*) Connoisseur coffees include sugar unless otherwise requested	**67p**

REFERENCES

Adda, J.: 1986, Fight against Academic Failure in Mathematics; In P. Damerow et. al. (Eds.) *Mathematics for All.* Document Series no. 20, UNESCO, Paris, pp.58-61.

Bernstein, B.: 1971, On the Classification and Framing of Educational Knowledge, Ch.2 in M. Young (Ed.) *Knowledge and Control.* London:Collier-Macmillan.

Caputo, J.: 1987, *Radical Hermeneutics: Repetition, Deconstruction and the Hermeneutic Project*, Bloomington and Indianapolis:Indiana University Press.

Carraher, D.: 1991, Mathematics in and out of School: A Selective Review of Studies from Brazil, Ch. 16 in M. Harris (Ed.) *Schools, Mathematics and Work* London: Falmer.

Cobb, P., Yackel, E. and Wood, T.: 1989, Young Children's Emotional Acts while Engaged in Mathematical Problem Solving Ch. 9 in D. McLeod and V. Adams (Eds.) *Affect and Mathematical Problem Solving*, New York:Springer-Verlag.

D' Ambrosio, U.: 1985, Ethnomathematics and its Place in the History and Pedagogy of Mathematics *For the Learning of Mathematics*, *5*, 1, 44-48; reprinted as ch.2 in M. Harris (Ed): 1991, *Schools, Mathematics and Work* London: Falmer.

Derrida, J.: 1978, *Edmund Husserl's Origin of Geometry: An Introduction*; transl. by J. Leavey, Lincoln and London:University of Nebraska Press.

Derrida, J.: 1988, *Limited Inc*, Evanston IL.:Northwestern University Press.

Evans, J.: 1991, Cognition, Affect, Context in Numerical Activity among Adults in F. Furinghetti (Ed.), *Proceedings of the Fifteenth International Conference of Group for Psychology of Mathematics Education* , Assisi, Italy, Vol. II pp. 33-39.

Evans, J.: 1993, *Adults and Numeracy*, Unpublished Ph.D Thesis, University of London Library.

Evans, J.: 1994, Gender and Mathematical Thinking: Myths, Discourse and Context as Positioning in Practices, In G. Hanna and B. Grevholm (Eds), *Study on Gender and Mathematics Education*, ICMI.

Evans J. and Tsatsaroni A.: 1993, Linking the Cognitive and the Affective: a Comparison of Models for Research, In *Proceedings of the Seventeenth International Conference of Group for Psychology of Mathematics Education*, Tsukuba, Japan

Foucault, M.: 1977, *Discipline and Punishment* transl. by A. Sheridan. Harmondsworth:Penguin.

Ginsburg, H. & Asmussen, K.: 1988, Hot Mathematics, in G. Saxe & M. Gearhart (Eds.).*Children's Mathematics*; New Directions for Child Development, no.41, San Francisco:Jossey-Bass, pp. 89-111.

Hamacher, W.: 1989, Lectio: de Man's Imperative, In L. Waters & W. Godzich (Eds.) *Reading De Man Reading*, Minnesota University Press pp. 171-201.

Hammersley, M. and Atkinson, P.: 1983, *Ethnography: Principles and Procedures*, London:Tavistock.

Henriques, J., Hollway, W., Urwin, C., Venn, C. and Walkerdine, V.: 1984, *Changing the Subject: Psychology, Social Regulation and Subjectivity*, London:Methuen.

Hollway, W.: 1989, *Subjectivity and Method in Psychology: Gender, Meaning and Science*, London:Sage.

Hunt, J.: 1989, *Psychoanalytic Aspects of Fieldwork*, London:Sage.

Jay, M.: 1982, Should Intellectual History Take a Linguistic Turn? Reflections on the Habermas - Gadamer Debate, In D. Lacapra and S.L. Kaplan (Eds.) *Modern European Intellectual History*, Ithaca NY:Cornell University Press.

Lacan, J.: 1977, *Ecrits*; transl. by A. Sheridan. London:Tavistock. (First published, 1966).

Laplanche, J. and Pontalis, J. B.: 1973, *The Language of Psychoanalysis*. The Institute of Psychoanalysis and Karnac Books.

Lave, J.: 1988, *Cognition in Practice: Mind, mathematics and culture in everyday life*, Cambridge:Cambridge University Press.

Maier, E.: 1980, Folk Mathematics, *Mathematics Teaching* 93.

Mellin-Olsen, S.: 1987, *The Politics of Mathematics Education*, Dordrecht:Reidel.

Nancy, J.-L. and Lacoue-Labarthe, P.: 1992, *The Title of the Letter: A Reading of Lacan*, State University of New York Press.

Newman, D., Griffin, P., & Cole, M.: 1989, *The Construction Zone: Working for Cognitive Change in School*, Cambridge:Cambridge University Press.

Nickson, M. and Lerman, S. (Eds.): 1992, *The Social Context of Mathematics Education: Theory and Practice*; Proceedings of the Group for Research into Social Perspectives of Mathematics Education. London:Southbank Press.

Noss, R., Brown, A., Dowling, P., Drake, P., Harris, M., Hoyles, C., and Mellin-Olsen, S. (Eds.): 1990, *Political Dimensions of Mathematics Education: Action and Critique; Proceedings of the First International Conference*, Dept. of Mathematics, Statistics and Computing, Institute of Education, University of London.
Reyes, L. and Stanic, G.: 1988, Race, Sex, Socioeconomic Status, and Mathematics *Journal for Research in Mathematics Education*, 19 1, 26-43.
Richardson, F., and Suinn, R.: 1972, The Mathematics Anxiety Rating Scale: Psychometric Data, *Journal of Counselling Psychology, 19,* 551-554.
Saxe, G.: 1991, *Culture and Cognitive Development: Studies in Mathematical Understanding*, Hillsdale NJ:Erlbaum.
Schoenfeld, A.: 1991, Mathematics as Sense-Making: An Informal Attack on the Unfortunate Divorce of Formal and Informal Mathematics, In J. Voss, D. Perkins and J. Segal (Eds.), *Informal Reasoning and Education*, Hillsdale NJ:Erlbaum.
Scribner, S.: 1984, Studying Working Intelligence; in B. Rogoff and J. Lave (Eds.) *Everyday Cognition*
Taylor, N.: 1990, Picking Up the Pieces: Mathematics Education in a Fragmenting World, In R. Noss et. al. (Eds.) pp. 235-242
Thom, M.: 1981, The Unconscious Structured as a Language, In C. MacCabe (Ed.), *The Talking Cure: Essays in Psychoanalysis and Language*, London:Macmillan.
Tobias, S.: 1978, *Overcoming Math Anxiety*, Boston:Houghton Mifflin.
Tsatsaroni, A.: 1991, Re-writing Professional Discourse, Unpublished Ph.D Thesis, University of London Library.
Tsatsaroni, A. and Evans, J.: 1994, Mathematics: the Problematical Notion of Closure, In P. Ernest (Ed.) *Mathematics, Education and Philosophy: an International Perspective*, London:Falmer.
Ulmer, G.: 1986, Sounding the Unconscious, In J. Leavey (Ed.) *Glassary*, Lincoln and London:University of Nebraska Press, pp. 23-129.
Walkerdine, V.: 1988, *The Mastery of Reason: Cognitive development and the production of rationality*, London:Routledge and Kegan Paul.
Walkerdine, V.: 1990, Difference, Cognition and Mathematics Education *For the Learning of Mathematics, 10,* 3, 51-55.
Walkerdine, V. and Girls and Mathematics Unit: 1989, *Counting Girls Out*, London:Virago

Jeff Evans,
School of Mathematics & Statistics,
Middlesex University,
Queensway,
Enfield,
Middx. EN3 4SF U.K.

Anna Tsatsaroni,
Department of Primary Education,
University of Patras,
Patras 26110,
GREECE

S. LERMAN

CHANGING FOCUS IN THE MATHEMATICS CLASSROOM

"It is the world of words which creates the world of things" [1]

The dominant paradigm within which research, testing, and theorising on the meaning of understanding are framed is a psychological one, which is generally interpreted as the study of the mind. Inevitably, therefore, it focuses on the individual. Recent ideas in mathematics education, drawn from discourse analysis (e.g. Pimm, 1990, this volume; Walkerdine, 1988, 1989) and social practice theory (e.g. Evans and Harris, 1991; Dowling, 1991), indicate that a consideration of social interactions does more than merely shed light on the understanding of the individual. Their critique suggests that understanding cannot be approached except through a focus on the social. The chapters in this book all pick up on a socio-cultural focus for the mathematics classroom and elaborate it from their own particular interests. In this chapter I will offer an interpretation of this shift in focus towards the socio-cultural, and suggest some activities for the mathematics classroom that may be seen to be fruitful in this context.

THE PSYCHOLOGICAL VIEW OF UNDERSTANDING

From the perspective of psychology, understanding is seen as a process that the individual undergoes, some kind of transition from not understanding a concept to understanding that concept. There are two elements in this transformation from a pre-understanding state to a state of understanding: the individual and the concept. The latter is assumed to be an objective, external element which in the educational environment is 'possessed' or held in some way, by the teacher. The teacher is seen to perform the intermediary role of introducing the concept, attempting to enable and facilitate its construction (or 'transmission to', or 'internalisation by' - no specific psychological paradigm is intended here) by the student, and measuring the outcome, that is whether the state of 'understanding' has been achieved. The former element, the individual, is largely a closed unit to the teacher, making the business of determining whether the transformation has taken place or not a very difficult one. Research in this area in education thus focuses either on interpretations of behaviour, or other student 'output', that might constitute evidence, using particular models, of conceptual development (van Hiele, 1986; Pirie and Kieran 1991; Steffe and Cobb, 1988), or on the conditions that will bring about 'understanding' (Hiebert and Carpenter, 1992).

What follows should be seen as somewhat of a caricature, and is based on the notion of a self-regulating 'normal' child. This image of learning as the

S. Lerman (ed.), Cultural Perspectives on the Mathematics Classroom, 191–213.

process of transformation from a pre-understanding state to a state of understanding determines the style and intention of the various stages of the educational process. It begins with assessing the initial cognitive state of the individual. It is followed by the prepared lesson, whose content is determined by research into hierarchies of mathematical knowledge and levels of difficulty of understanding, which are determined by large-scale tests using well-established quantitative procedures. Great emphasis is placed on how mathematical concepts and skills are explained by the teacher. Research on affective factors influence the classroom setting, the appearance of the text, etc. On-going assessment enables the identification of misconceptions, misunderstandings or partial understandings. These can be corrected by suitable re-explanations, reinforcement materials and so on. The final stage is the test that measures the successful acquisition of the particular skill or concept that is interpreted as understanding, and again that test will have been trialled over large groups of students with well-established techniques.

The intention of this caricature is to highlight the manner in which the paradigm dominates and determines the whole structure of the educational process. It is the essentially private nature of the psychological interpretation of understanding that pervades the educational process from this perspective. The task of the teacher would be enhanced by the existence of a light bulb attached to one's head, which lit when true understanding took place. In the absence of such an evolutionary or technological development, teachers are left to try as best they can to recognise when that process has occurred and, given recent developments in British education, tick the appropriate box to register that attainment target for the student.

It is the case, I am suggesting, that our assumptions about the nature of mathematics (Lerman, 1983, 1990) and the nature of the learning process become self-fulfilling. We interpret, theorise, teach, test, assume, expect, measure and thus confirm our initial expectations of children. Walkerdine, following Foucault, refers to this as a 'regime of truth' (Walkerdine, 1988 p. 31).

At the same time, there is an assumption, usually implicit, that we all arrive at the same understanding of particular *things*, because of the essential nature of those *things*, whether physical objects, such as tables, and unicorns, or mental objects such as circles, or 'three'. A critical analysis, therefore, needs to work on both aspects of the understanding process in the psychological paradigm, namely the notion of the individual, and also the nature of knowledge.

The development of interest in Radical Constructivism over the last decade or so has been driven by a radical position on both the individual and the nature of knowledge. Since the individual constructs knowledge for her or himself, what constitutes 'knowing' can only be interpreted as that which the individual conceptualises. We do not come to know a real world. Each one of us constructs a personal reality, based on our experiences, which is modified, or not (see below), by interactions with one's environment:

Sensory-motor material, graphic representations, and talk can provide occasions for the abstraction of mathematical operations, but they cannot convey them ready-made to the student. (von Glasersfeld 1992)

Radical Constructivism takes the psychological focus on the individual to its ultimate extreme, leading some to talk of working with children as "in many ways like working with foreigners with whom one has only fragments of a language in common (Steffe and von Glasersfeld, quoted in von Glasersfeld 1991 p. 22). This position is discussed further below.

CRITIQUE: THE INDIVIDUAL

As I have said, in the psychological paradigm, the major focus of attention is on how the individual functions and acquires knowledge. The individual may be thought of as the *subject* as in a grammatical sense, that is taking the position of the performer of acts, in front of the verbs. There is a sense in which the subject is autonomous, decisive, independent and acts upon things and indeed people. However, an analysis of the complexities of the social setting of the classroom, the language, culture, and power relations, suggest that the *subject* is simultaneously the apparently autonomous self-constituted individual who articulates a discourse and also the *object of* that discourse. The teacher is also one of the *objects* of the collection of actions, in a sense coming after the verbs; put another way, she or he is *subjected* by the discourse.

Discursive practices are domains of social interaction, by which is meant the interactions of language, power relations, knowledge and social practices. These interactions will be elaborated upon below, but to summarise here: language is specific to particular social practices and is associated with power as knowledge and knowledge as power; language structures what we can talk about, as the quote from Lacan that heads this chapter implies; and

. . . there is no knowledge without a particular discursive practice; and any ‹ iscursive practice may be defined by the knowledge that it forms. (Foucault, dioted in Said 1974)

Discursive practices are not clearly bounded, they are continually changing, and one switches from one discursive practice to another. In a series of interviews, Evans (1991, this volume) asks mathematical questions set in different social contexts, and attempts to identify the discursive practice that is called up by the question in its context, for a particular person. He criticises the simplistic notion that giving real world contexts for mathematical concepts provides 'meaning' for students, a 'meaning' that supposedly exists in some absolute sense and is illustrated by or modelled in that real world context. He also criticises the notion that affect and cognition are separable aspects of behaviour. Vygotsky points to the same issue (1986):

> When we approach the problem of the interrelation between thought and language and other aspects of mind, the first question that arises is that of intellect and affect. Their separation as subjects of study is a major weakness of traditional psychology . . . (p. 10)

It has often been said that one of the main alienating features of school mathematics is that it appears to be about *nothing*. Thus making it about *something* is intended to make connections for students and enable meanings to be negotiated because a meaningful context is provided. Pimm (1990) discusses this in terms of metaphoric links on the one hand and metonymic chains on the other. He suggests that mathematical activity is characteristically metonymic, so that the constant reference to metaphors can be misleading. It may be that our efforts to make mathematics more user-friendly drives the search for metaphors "Negative numbers are like debts in the bank". The issue of making mathematics about something is developed in Frankenstein (1990) with the intention of problematising the role of mathematics in society, but I want to carry the critique deeper into the mythology of mathematics. To operate with the notion that the teacher can provide a meaningful context for the students, in some general neutral way, such that they will be able to abstract the mathematical ideas more successfully than by teaching the skills in themselves, as they are, requires the further analysis that Evans and others are engaged in.

> . . . mathematical activity and context cannot be neatly separated: they are part of the same whole, and both are shaped and made meaningful by the larger activity or practice(s) of the subject. The latter are themselves shaped and made meaningful by language. (Evans, 1988 p. 297)

Clearly, Frankenstein is saying more than this, that the problems themselves are intended to generate questions that go beyond the learning of mathematics in a classroom, towards a recognition of mathematics as power, and that students have that power within their grasp too. 'Context', however, is not so simply generalisable, since the culture of the individual is such a complex mixture of class, ethnicity, gender, emotions, family dynamics, classroom dynamics and so on. What is meaningful for one may not be for another, and what a signifier calls up may well be different too. At the same time, what I have called here 'culture' is conceived through, expressed in, and bounded by communication, and in particular language, and not some private uncontactable world. Any reconstruction of mathematics teaching and learning needs to take into account the embedding of learning and mathematics in students' cultural worlds. That is to say, just as the teacher cannot do the learning for the students, so s/he cannot create the meanings or make contexts relevant. The teacher's task is to attempt to create an environment in which students will draw into the classroom setting whatever meanings are called up for them, but also one in which they will confront and engage with other meanings that, in our case, mathematics calls up.

Walkerdine (1988) offers a number of illustrations of classroom talk, in which the teacher does not necessarily recognise that discursive practices other than the classroom mathematical one are called up for the child, often with strong emotional connections. Small wonder that there are misunderstandings, disagreements, loss of confidence leading to 'failure' in school mathematics. Yet another illustration, a very vivid one, is given by Walkerdine et al (1989) in a description of a classroom incident involving four and five year olds, where the discourse dramatically and suddenly changes, and with it the language and power relations, thus highlighting that switch:

Annie takes a piece of Lego to add to a construction she is building. Terry tries to take it away from her to use himself . . . The teacher tells him to stop and Sean tries to mess up another child's construction. The teacher tells him to stop. Then Sean says: 'Get out of it, Miss Baxter paxter.'
TERRY: Get out of it, knickers Miss Baxter.
SEAN: Get out of it, Miss Baxter paxter.
TERRY: Get out of it, Miss Baxter the knickers paxter knickers, bum.
SEAN: Knickers, shit, bum.
MISS BAXTER: Sean, that's enough. You're being silly.
SEAN: Miss Baxter, knickers, show your knickers.
TERRY: Miss Baxter, show off your bum (they giggle).
MISS BAXTER: I think you're being very silly.
TERRY: Shit, Miss Baxter, shit Miss Baxter.
SEAN: Miss Baxter, show your knickers your bum off.
SEAN: Take all your clothes off, your bra off. . .
(p. 65-66)

Walkerdine adds: "People who have read this transcript have been surprised and shocked to find such young children making explicit sexual references and having so much power over the teacher." (p. 66). As I have said, power relations and positions are constructed by discursive practices and this transcript shows how those relations have been turned upside down, with the children suddenly in control, by virtue of the shift to the sexist discourse of the male children to the female teacher. This is not to suggest that the transformation is *intended* by the boys; some slippage has taken place, perhaps through rhyming. Language is associated with power, resulting in positioning of the actors in the discourse. In this case, the sexist language places the boys in the position of power.

These examples are intended to demonstrate where the focus of study shifts, in this critique. An analysis of how language operates within discursive practices indicates the way in which knowledge, understanding, meaning and ourselves as subjects are constituted in and bounded by our language, and that language is specific to discursive practices. The notion of the autonomous individual, coming face to face with external concepts, as physical or mental objects or things, and acquiring an 'understanding' in a private process is not adequate.

A competing view of 'understanding' is one that sees the process as social, rather than psychological. That is to say, the individual does not go through some transformation from *pre* to *post* state of understanding in isolation, acquiring a given well-formed concept in some hidden way. It is necessarily an interaction of the child and her multitude of meanings, through language, with others and with experiences. Sometimes the understandings do not fit with those of others, including the teacher, and only talk, discussion, suggestions and conjectures and refutations, or shifts of thought through resonance, enable further growth.

The analysis of context, as discussed above, is not only a matter of attempting to embed mathematics in situations that may be 'relevant' or 'meaningful' for students, but also an analysis of the discursive practices within which mathematical terms and concepts appear, both for students in their own experiences and for the mathematical community.

I have attempted to suggest, when conceptualising the 'individual' in the process we call understanding, that the focus is on the social interactions, the language and its meaning for the individual and the groups within which she acts. I will turn now to the other aspect, namely knowledge.

CRITIQUE: OBJECTIVE KNOWLEDGE

Traditionally, psychology is not much interested in the nature of knowledge. Concepts are taken as givens, and the concern is with how the individual can acquire that knowledge and retain it. Piaget, however, was a philosopher first, and his concern with psychology arose from his premise that one can only approach knowledge through investigating the manner of its acquisition. In his view, knowledge is neither empirical nor a priori in a platonic sense. Rather he believes that "cognition produces conceptual structures by *reflective abstraction* from material that is available within the system and from the operations carried out with that material" (von Glasersfeld, 1991, p. 16). This is a very individualistic interpretation, and Piagetians have different ways of avoiding this solipsism. Piaget's approach was to fall back on the *necessity* of mathematical structures, and the assumption that we will all arrive at the same mathematics. Von Glasersfeld talks in terms of the fit between the individual's concepts "formed out of elements from his or her subjective field of experience" (ibid p. 23) and those of other speakers of the language. I will return to von Glasersfeld's radical constructivism below, but his interpretation does not appear to be guaranteed to lead one out of solipsism.

Consider the well-known example from philosophical discussions, namely what is meant by saying that a child has learned and understands the concept 'hat'. This comes about by pointing out instances of 'hat', objects that have the use implied by that term. When the child points to a hat and says "Hat", we confirm that this is correct. When the child points to a tea-cosy and says "Hat" we have to explain that this object has a different use, and is not a hat. Where then is the concept 'hat'? It is in the use, according to the public, objective notion of 'hat', that we can apply the word 'understand', and it has no application without this public connection. (Of course, someone, initially called eccentric, and later perhaps a person who sets a new trend, may put the tea-cosy on their head, and call it a hat, and our concept will have to undergo a public change.) Wittgenstein says (1958):

> Every sign *by itself* seems *dead*. *What* gives it life? In use it is alive. Is life breathed into it there? - or is the *use* its life? (Remark 432)

Two main points arise from this example, firstly that the concept is socially determined, and secondly that we do not expect that the understanding of the child comes about in some hidden way. It is clear that the concept will be acquired, or perhaps 'appropriated' might be a more suitable term, through looking at examples of the use of the term, and inviting the child to test for her/himself, against the social conventions surrounding the word, the newly accepted concept.

Another example is to be seen in what Wittgenstein says about understanding (Wittgenstein 1974):

> Do I understand the word 'perhaps'? - And how do I judge whether I do? Well, something like this: I know how it's used, I can explain its use to somebody, say by describing it in made-up cases. I can describe the occasions of its use, its position in sentences, the intonation it has in speech. - Of course this only means that 'I understand the word "perhaps"' comes to the same as: 'I know how it is used etc.'; not that I try hard to call to mind its entire application in order to answer the question whether I understand the word. (p. 64)

The concepts used above, 'perhaps' and 'hat' may appear more easily characterised as social in nature than mathematical concepts, and therefore their acquisition can be more easily described in social processes. This raises questions about the nature of mathematical concepts, and whether they are any different from 'hat' and 'perhaps', possibly because of a feeling of the universality, or the certainty of mathematical ideas.

Are mathematical concepts any different? They are abstract certainly, and are therefore encountered in ways other than concepts such as 'hat', but I would argue that this does not make them any the less social in nature than 'hat' or 'perhaps'.

There are anthropological studies such as Pinxten (1988, this volume), and Douglas (1973) that indicate how mathematical notions function in different societies, including Western society, and studies such as Grabiner's (1986) that demonstrate how notions of 'proof' have changed over time, and have served different purposes. David Bloor's work (1976) has extended sociological analyses into interpretations of scientific and mathematical knowledge itself, previously taboo areas for sociology, and in a previous article I wrote (Lerman 1989):

> (Bloor) takes the views of Frege, one of the major figures in presenting the neo-platonist image of mathematics, to illustrate the social nature of mathematical knowledge. Frege uses the equator, the axis of the earth and the centre of mass of the solar system as examples of objective but non-physical entities, which is how he characterises mathematical existence. Bloor points out how these are just social constructions, invented by people to function as structuring and ordering concepts. In the Ptolemaic system, the centre of the earth served the same role, as the centre of the universe which consisted of concentric circles around the circular earth. Bloor comments that Frege would be as horrified of 'sociologism' as he was by 'psychologism', as he termed Mill's empiricist philosophy, the latter being the idea that concepts gain meaning in the individual mind, and the former in the social mind, as it were. (p. 219)

Eco (1976) captures the same essential perspective when he writes:

> Every attempt to establish what the referent of a sign is forces us to define the referent in terms of an abstract entity which moreover is only a cultural convention." (p. 66)

Rotman attacks the separation of mathematical objects from their descriptions in a (1988) paper, in which he writes:

> . . . what mathematicians think they are doing - using mathematical language as a transparent medium for describing a world of pre-semiotic reality - is semiotically alienated from what they are, according to the present account, doing - namely, creating that reality through the very language which claims to 'describe' it. (p. 30)

These objects in mathematics, then, are objective in an intersubjective sense, agreed, useful, long-lasting but potentially changeable. Concepts are socially determined, and thus socially acquired. Since concepts gain their meaning in their use, the acquisition of a concept, or 'understanding', can be interpreted as that of an individual coming to share in that meaning through negotiation and discussion, and in a sense, conjecture and refutation or corroboration. The language creates reality and then describes that creation, and it can appear as though the description is that of a timeless, fixed reality. There is a sense in which the 'grammar' of the created reality takes on a life of its own, just as the rules of chess create almost endless possibilities. Yet there are always new situations. In tennis, were it possible for a person to throw the ball into the air to a height of 100 metres when about to serve, doubtless a new rule would have to be invented (Shankar 1987).

A RECONSTRUCTION: THE INDIVIDUAL AND KNOWLEDGE

In this discussion I have separated the individual from knowledge, but only in order to demonstrate that they are not separate. Just as the subject is constituted in discursive practices, so too are concepts identified and characterised by use which is again constituted within social practices. The psychological paradigm treats the individual as autonomous, able to relate to the reality of the outside world or not, and to 'understand' external concepts or not. But, I have argued, the 'individual' cannot be divorced from the social, and neither can 'concepts'. Indeed, focusing on the social construction of subject and of knowledge brings the two together.

Returning briefly to the radical constructivist ideas of von Glasersfeld and others; whilst they write in terms of reality as a construction rather than something external and absolute that is passively received, their focus is still the individual. They do not deny that people interact; on the contrary social interactions are one of the ways that the individual tests out the viability of her or his reality. The sense in which 'social' is interpreted, however, from that individualistic focus, is in a secondary, almost unnecessary role.

> The experiential environment in which an individual's constructs and schemes must prove viable is always a *social* environment as well as a *physical* one. Though one's concepts, one's ways of operating, and one's knowledge cannot be constructed by any other subject than oneself, it is their viability, their adequate functioning in one's physical and social environment, that furnishes the key to the solidification of the individual's experiential reality. (von Glasersfeld 1991, p. 20-21)

One has a sense, from this paragraph, that von Glasersfeld pictures the individual as creating concepts independently and then comparing them with the outside world, both social and physical. Indeed in relation to language Piaget writes (1969) that it plays:

> . . . a central role in the formation of thinking only in so far as language is one of the manifestations of the symbolic function. The development of the symbolic function in turn is dominated by intelligence in its total functioning. (p. 126)

There is a clear sense of every individual having a language and concepts of her/his own, which become adapted to others through interaction. But the judgements of 'viability' and 'adequate functioning' are not objective and universal criteria, and they are themselves not passively received from outside

(Lerman 1993). Indeed "... we come to see knowledge and competence as products of the individual's conceptual organization of the individual's experience ... " (von Glasersfeld, 1983 p. 66) There is no certainty here, as they would happily agree. How then is there any communication between people? Is there not an infinite process of the teacher's construction of the child's construction of the teacher's ... etc? There is no connection, the worlds are separate, and finally incommensurable.

For Piaget, and hence for von Glasersfeld and the Radical Constructivists, thought precedes language, and without some 'universals', such as platonic forms, or an external world which forces its nature upon us, we are imprisoned in our own constructed environment. It is the extreme of incommensurability, since even the existence, or otherwise, of a 'fit' is determined by the autonomous self.

But the shift of focus from the individual to the social locates language and communication differently (Weedon 1987):

> Language is not transparent ... it is not expressive and does not label a 'real' world. Meanings do not exist prior to their articulation in language and language is not an abstract system, but is always socially and historically situated in discourses.

There is thus an important difference between the notion that the individual thinks first and uses language to judge a 'fit' with reality, even social realities, and the notion that thoughts are constructed in language which is culturally and historically situated. In the latter sense of language, Wittgenstein has argued (e.g. 1958) that there is no problem of private languages. As in the discussion above of the concept 'hat' there is no *life* to the concept before, after or without its *use* . 'Hat' is as much a social construction as 'equator', 'unicorn', 'three', or 'polyhedron'. A private understanding, to be subsequently compared with something outside in order to judge a 'fit', is a perception that belongs to the individualistic, psychological paradigm.

In the following section, the discussion moves from a theoretical perspective to a focus on activities in the classroom that may be useful in deconstructing the image of mathematics.

DECONSTRUCTING AND RECONSTRUCTING MATHEMATICS IN THE CLASSROOM

A shift of focus from the individual to social interactions arises, as is the nature of paradigm shifts, both from critique of the dominant paradigm and from elaborations of the potential richness of the alternative perspective.

Action and Critique are to be united in the providing of resources for the deconstruction of mechanisms of social and cultural reproduction. (Dowling, 1990 p. 81)

The research on discourse analysis and social practice theory described above are attempts at both critique and reconstruction, although the latter is still in need of much elaboration and will certainly take new forms. Brown and Dowling (1989), for instance, propose what they call a "research-based" approach:

Our method has been to propose a question - say "who does the best at school?" - as the basis for a research project. (p. 37)

In the final part of this chapter, I propose some activities that identify potential sites for action and critique in the classroom, namely: the school mathematics textbook; the 'certain' image of mathematics, and the image of the steady progress of mathematical knowledge as 'revealed' by the history of mathematics.

School Textbooks

In recent years, some of the well-established school mathematics texts have been revised, in an attempt to remove gender and race bias, and new series have taken the issue of equal opportunities into account in their writing. Thus, in Britain, the workcards in the individualised learning scheme SMILE have undergone this process, revisions such as "SMP 11-16" - the latest version of the most commonly used texts in secondary schools, the School Mathematics Project - and new series such as National Mathematics Project (NMP) attempt to show equal numbers of boys and girls, and a racial mix, in their pictures, and examples generally try to avoid stereotypes, such as mothers shopping and fathers at work.

There remain, however, strong social messages conveyed in texts. The SMP 11-16 books for students aged 14 and above are divided into sets for different 'abilities', the Green series G1 to G8 for the lowest ability and Y1 to Y5 for the highest ability. The appendix shows extracts from these two series, concerned with Income Tax, the first from the low ability text G8 and the second from the high ability text Y5:

One immediately notices that the lower ability text is dominated by visual images concerned with unemployment and leaving school, whereas the high ability text has a drawing of white male civil servants dividing up piles of money into categories. Further examination reveals that the high ability text contains a table showing the tax rates for bands of income, right up to earnings over £40,000 per year, but the low ability text does not give any information about different rates, merely the lowest tax band. Indeed, looking at the examples used in the two texts, the lower ability one refers to earnings up to only £9000, whereas the high ability text starts with a question about an income of £50,000.

There are many assumptions here, centered around the idea that low ability in mathematics (whatever that may mean) correlates with low ability and intelligence in general, with leaving school at 16 and low expectation in careers and salaries, with low reading levels (note the size of the print, length of sentences, choice of visuals), etc.

This style of critique of texts is not new (see e.g. Maxwell 1985), but is a continuation of the critiques centred on gender and ethnicity of some years ago, and reveals that social messages through mathematics texts and examples run deep. One could suggest that the authors go back to the drawing board, and try harder.

However, I want to consider a different strategy, namely that of offering both texts, for comparison, to both 'ability' levels of students? One could invite students to note the messages of the two texts, and discuss why these assumptions are made. Ways of validating or invalidating such assumptions could be suggested and tested by the students, etc.

I am suggesting here an extension of the work proposed in some recent books (e.g. Frankenstein 1990). In that book she asks, for example, why should we use studies of which programmes on television students watch, for learning statistics, when we could encourage students to problematise the information fed to us by the media and government agencies. In my example I am suggesting that we invite students to challenge the source of objective mathematical knowledge in the school, namely the mathematics textbook itself. I recently presented the idea described, namely inviting students to look at the same topic in different school books for different abilities, to an inservice course of teachers, and again the reaction was one of shock. One teacher said he couldn't do that, it would be too dangerous. This is quite valid, challenging the role of texts as the source of established objective knowledge is dangerous. Suddenly one recognises that there are textbook writers, they are people, with their own values, opinions, beliefs and ideas about mathematics - who it is for, what people should learn, etc. The teacher of mathematics is identified with the textbook writer as part of the authority of mathematics, and thus the challenge to the textbook is a challenge to the mathematics teacher. Further, the published word is a particularly powerful medium for mathematics, since publication of mathematical work provides legitimation and ultimately career and status. The books may not be made of stone, but the words have the same lasting character as if they were. However, as Davis says, (Davis 1986):

> A past editor of the *Mathematical Reviews* once told me - somewhat in jest - that 50% of all mathematics papers printed are flawed. (p. 174)

Texts convey the content of mathematics to be learned: the order or hierarchy of the ideas, both a mathematical hierarchy and a developmental hierarchy; messages about the nature of mathematics; messages about society, how it functions and what is appropriate; messages about the individual student and his/her function in society, etc. The intention of this present illustration is to problematise the established notions of mathematics for ourselves as teachers and for students. Analysing school texts, and examining the historical development of the function that the text serves in schools will certainly play a part in deconstructing the mythology of the mathematical absolute.

The 'Certainty' of Addition

In any discussion challenging the objectivity and certainty of mathematics, the reply is often that there is, for example, only one outcome for the addition or multiplication of two numbers. Philip Davis questions this in a fundamental way (Davis 1986), and I have attempted(!) to reproduce his example below:

PROBLEM: Given
A=11777777711171717171777171171711111177717177711771177171717
17177717177717171717177711171711111171777111717171111117171771
71
B=777771711711177777771111111177171717111777771717777711171711
11171711717177711111117171777777771117171777711117771171777771
Find A + B.
The numbers A and B cannot be reproduced with perfect fidelity, let alone added." (page 171)

The surprise for the reader here is that the difficulties of practice, namely the extent to which we can be sure that any person can repeat even the writing down of these figures with certainty, or fidelity as Davis describes it, interfere epistemologically. Can we say that there **is** an answer to this question? If so, where is it, and how can we know? It is no answer to suggest that human eyes are fallible, and so we should give the problem to a computer, since any scanning process by a computer places us in exactly the same position. How do we know if the computer has 'read' the problem correctly, and for that matter whether it has carried out the addition correctly? We have no certain answer by which to check or verify, since for arithmetic, by 'certain answer' we usually mean verifiable by hand by an established algorithm.

To set students on this problem, and to discuss the implications for images of mathematics, would again be, I suggest, problematising the given and accepted notions of mathematics. Following Davis's example, what are we left with, as far as certainty in mathematics goes? Perhaps no more than the certainty of a game of chess, that has its rules, or its grammar, and provided one follows the rules, and stays within that limited context, things can proceed in verifiable procedures. So too in mathematics, we can verify a calculation based on a surveyable list of numbers, but this is quite limited, as the example shows. To quote Davis (1986) again he says, somewhat tongue in cheek:

The sum 12345 + 54321 is not 66666. It is not a number. It is a probability distribution of possible answers in which 66666 is the odds-on favourite. (page 171)

Einstein wrote, in 1921:

Insofar as the propositions of mathematics give an account of reality they are not certain; and insofar as they are certain they do not describe reality.

However the critique that is indicated in the example of surveyability questions what is meant by certainty itself, in particular in relation to mathematics. Surveyability, individual interpretation, abstraction from 'reality' to the point of 'mathematical fiction' are characteristics of mathematical certainty that are not generally recognised, nor indeed mentioned. This is not to claim that proof and certainty, deduction and induction have no relevance for mathematics, but to attack their reification and place them in the social domain.

History of Mathematics

The following are extracts from a recent study in the history of mathematics (Henley 1990):

The first recorded appearance of negative numbers seems to have been from 1600 BC and can be found in Babylonian clay tablets at Yale University"
"For instance, solve the equations
$xy = 600$ and $150(x-y) - (x+y)^2 = -1000$"
. . . From this selection of references we find conflicting views. Kline [1972] flatly denies that the Babylonians had any concept of negative numbers . . .

Later he quotes from a British mathematician, William Frend, who wrote in 1796:

. . . to attempt to take (a number) away from a number less than itself is ridiculous . . . This is all jargon, at which common sense recoils.

The history of mathematics is generally seen as a positivist programme of the demonstration that the world is ultimately written in the language of mathematics, as Galileo thought, and history is the developing story of the revelation of this truth. When people are faced with discoveries, such as that negative numbers are numbers, it is supposed to be as if it was obvious all the time.

Here however we see firstly that negative numbers appear to have been quite acceptable in 1600 BC yet unacceptable as late as 1800 AD (and later, according to Henley's research).

An anecdote:
A student reported to me that he had been brought up in a monastic school in Scotland, where to the age of 18 he was told that larger quantities can never be taken away from smaller quantities, and to attempt to do so was the work of the devil.

Secondly, we see that the history of mathematics itself is subject to interpretation. Kline clearly interprets the mathematical attainments of Babylonian culture in one way, whereas Henley points out a quite different image. The issue of the dominance of the vision that all Western thought is descended from the Greeks, who themselves broke with previous civilizations in their approach to knowledge, is treated more fully in Bernal (1987). This kind of study of the history of mathematics, quite distinct from the spicing up of the regular curriculum with snippets of information, can have the effect of demystifying the positivist view of the history of mathematics and hence of the nature of mathematics itself (Lerman 1992).

CONCLUSION

In this chapter I have attempted to contribute to the growing critique of the individualistic focus of educational psychology and the shift to a focus on the social as providing the sites of the construction of the *subject* and of knowledge. In the first part of the chapter I examined some of the consequences and identified some of the hidden assumptions of the individualistic paradigm. I attempted to demonstrate places where the critique bites strongly and some of the alternatives for thinking about the mathematics classroom that are offered by a focus on language and discursive practices. In the second part I proposed some activities for the mathematics classroom that identify and critique some of the mechanisms of perpetuation of a particular image of mathematics. It may be that these activities are more appropriate for teachers, but in a sense this is immaterial. What takes place in the mathematics classroom is at least strongly influenced, if not determined, by the beliefs of the teacher (Lerman, 1990, Scott-Hodgetts and Lerman, 1990).

There are a growing number of research programmes whose stimulus is the shift in focus from the individual's 'understanding' to the social nature of thought, knowledge-creation and learning, and whose agenda is the critique and re-construction of mathematics education. Some have been mentioned, e.g. Evans (1991) and Dowling (1991). Others, from different aspects, include Sutherland's (1992) work on the development of algebraic symbolism as communication and Morgan (1992) on childrens' writing as communication of mathematical meaning. The "world of words" is indeed a rich resource for research in mathematics education.

NOTES

1. from "Ecrits 1" J. Lacan, Seuil, Paris 1966 p. 155

APPENDIX

SMP BOOK G8 PAGE 42

D Tax

The government needs money to spend on the National Health Service, Defence, Social Security and so on.
It gets the money from taxes.

There are lots of different taxes, like income tax and VAT (value added tax).

Income tax is what you pay when you earn money.
You pay VAT when you buy things.

The rates of tax change from time to time.

D1 In 1974, the VAT on HiFi's was 8%.

(a) What is 8% of £250?
(To find 8%, multiply by 0·08.)

(b) How much did this HiFi cost in 1974?

(c) In 1975, the VAT rate on HiFi's was 25%.
How much would this HiFi have cost in 1975?

D2 In 1987 the VAT rate was 15%.
How much would a HiFi have cost in 1987 if the price was '£250 + VAT'?

42

SMP BOOK G8 PAGE 43

When you have a job you are paid wages. Usually your employer takes income tax off your wages. (This is called PAYE – Pay As You Earn.)

Everyone is allowed to earn some money each year without paying tax. In 1987, a single person could earn £2425 in a year without paying tax.

If a single person earned more than £2425, he or she had to pay income tax.

Personal allowances 87/88	
Single person	£2425
Married person	£3795
Additional allowance for single parents	£1370
Widows bereavement allowance	£1370
Age allowance	

Here is an example.

In this year, Salma earned £8600 before tax. Salma is single, so she can earn £2425 without paying tax.

So she has to pay income tax on £6175.

~~Married~~/single	P51Ua
name Salma Jadiq	87/88
Amount earned 87/88	£8600·00
Personal allowance	£2425·00
Taxable income	£6175·00
Income tax at 27%	£

D3 In 1987/88, income tax was 27%.

(a) How much tax will Salma pay this year?

(b) Salma earned £8600 before paying tax. How much money was she left with after paying tax?

D4 (a) Copy this form for Darren Smith.

(b) Complete the parts which are blank.

(c) How much money was Darren left with after paying his income tax?

~~Married~~/Single	P5iUa
Name Darren Smith	87/88
Amount earned 87/88	£4300·00
Personal allowance	£
Taxable income	£
Income tax at 27%	£

D5 Faith sells ice-creams.

On average, she earns £180 a week, before tax.

(a) How much does she earn in a year?

(b) How much tax would she pay in 1987/88?

43

SMP BOOK Y5 Page 18

Money matters: income tax

The government spends a lot of money each year, on such things as education, health, defence, and so on. Much of this money comes from **income tax**. The amount which anyone has to pay depends on their income – the amount of money they have coming in.

People do not pay tax on the whole of their income. There is part of their income which they are allowed to keep without any tax being deducted from it. This part is called their **allowances**. The size of the allowances depends on a whole variety of things.

Find out

It is possible for someone to have an income and not pay any income tax.

Find out what is the most you can earn without having to pay tax.

The **rate** at which you pay tax depends on how much you earn.
Your taxable income (your income minus allowances) is 'sliced' into **tax bands**. Each band is taxed at a different rate.

SMP BOOK Y5 PAGE 19

This diagram shows the tax bands and tax rates in 1986–87.

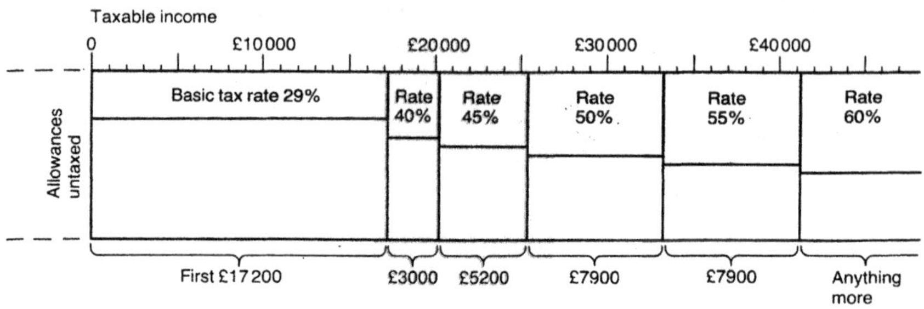

Find out

Find out what the tax bands are at the moment.

Work out how much tax is paid altogether by someone with a taxable income of £50 000.

Tax facts

- The 'tax year' (the year for which your income is calculated for tax purposes) runs from 6th April in one year to 5th April in the next year.

  Before 1752, rents and taxes of all kinds were calculated up to the end of the first quarter of the year – 25th March, called 'Lady Day'.

  In 1752 an adjustment was made to the calendar, so that 2nd September was followed by 14th September. So from Lady Day 1752 to Lady Day 1753 there were only 354 days.
  The Treasury accounting system could not cope with that, so from 1753 taxes were collected 11 days later, on 5th April!

  When income tax was first introduced in 1799, 5th April was the date on which it was collected.

- For nearly 30 years in the 19th century there was no income tax. It started again in 1842. The rate was 3%!

- In 1875 the rate of income tax was down to 0·83%.

- The highest the (basic) rate has ever been was during the Second World War, when it was 50%.

- There are two small islands around the British Isles where no income tax is paid by the inhabitants – Lundy Island and Sark.

REFERENCES

Bernal, M.: 1987, *Black Athena - The Egyptian Origin of Greek Civilisation* London:Free Association Books.

Bloor, D.: 1976, *Knowledge and Social Imagery* London:Routledge and Kegan Paul.

Brown, A. and Dowling, P.: 1989, *Towards a Critical Alternative to Internationalism and Monoculturalism in Mathematics* Education Centre for Multicultural Education Working Paper No. 10, Institute of Education, University of London.

Davis, P. J.: 1986, Fidelity in Mathematical Discourse: Is One and One Really Two? In *New Directions in the Philosophy of Mathematics*, T. Tymoczko (Ed.), Boston:Birkhauser.

Douglas, M.: 1973, *Natural Symbols: Explorations in Cosmology* Harmondsworth:Penguin.

Dowling, P.: 1990, Some Notes Towards a Theoretical Model for Reproduction, Action and Critique In R. Noss et al (Eds.) *Political Dimensions of Mathematics Education: Action and Critique*, Institute of Education, University of London p. 76-84.

Dowling, P.: 1991, The Contextualising of Mathematics: Towards a Theoretical Map In M. Harris (Ed.) *Schools Mathematics and Work* Basingstoke:Falmer p. 93-120.

Eco, U.: 1976, *A Theory of Semiotics* London:Macmillan.

Einstein, A.: 1921, (quoted in) M. Kline *Mathematics: The Loss of Certainty* 1980 Oxford:Oxford University Press p. 97.

Evans, J. and Harris, M.: 1991, Theories of Practice In M. Harris (Ed.) *Schools Mathematics and Work* Basingstoke:Falmer p. 202-210.

Evans, J.: 1988, Contexts and Performance in numerical activity among adults in *Proceedings of Twelfth Annual Conference of the International Group for the Psychology of Mathematics Education*, Hungary, Vol. 1, p. 296-303.

Evans, J.: 1991, *Cognition, Affect, Context in Numerical Activity Among Adults: The Contribution of Ideas from Post-structuralism* unpublished paper presented at Ninth Meeting of Group for Research into Social Perspectives of Mathematics Education, South Bank Polytechnic, London.

Frankenstein, M.: 1990, *Relearning Maths: A Different Third R - Radical Maths*, London:Free Association Books.

Grabiner, J.: 1986, Is Mathematical Truth Time-Dependent? In *New Directions in the Philosophy of Mathematics*, T. Tymoczko (Ed.), Boston:Birkhauser.

Henley, A.: 1990, *The History of Mathematics as a Pedagogical Aid, with Special Reference to Negative Numbers* unpublished MSc Dissertation, South Bank Polytechnic, London.

Hiebert, J. and Carpenter, T. P.: 1992, Learning and Teaching with Understanding, In D. A. Grouws (Ed.) *Handbook of Research on Mathematics Teaching and Learning* New York:Macmillan.

Lerman, S.: 1983, Problem-solving or Knowledge-centred: the Influence of Philosophy on Mathematics Teaching *International Journal of Mathematical Education in Science and Technology*, Vol. 14, No. 1, 59-66.

Lerman, S.: 1989, Constructivism, Mathematics and Mathematics Education *Educational Studies in Mathematics*, Vol. 20, p. 211-223.

Lerman, S.: 1990, Alternative Perspectives of the Nature of Mathematics and their Influence on the Teaching of Mathematics *British Educational Research Journal*, Vol. 16, No. 1 p. 53-61.

Lerman, S.: 1992, Critical/Humanistic Mathematics Education: A Role for History? in *Multicultural Teaching*, Vol. 11, No 1, p. 24-26.

Lerman, S.: 1993 The Position of the Individual in Radical Constructivism: In Search of the Subject, In J. Malone & P. Taylor (Eds.), *Constructivist Interpretations of Teaching and Learning Mathematics* Perth:Curtin University of Technology.

Maxwell, J.: 1985, Hidden Messages *Mathematics Teaching* No. 111, June p. 18-19.

Morgan, C.: 1992, Written Mathematical Communication: The Child's Perspective, unpublished paper, South Bank University, London.

Piaget, J.: 1969, Language and Intellectual Operations, In H.G. Furth (Ed.), *Piaget and Knowledge*, New Jersey:Prentice-Hall.

Pimm, D.: 1990, Certain Metonymic Aspects of Mathematical Discourse, In *Proceedings of Fourteenth Annual Conference of the International Group for the Psychology of Mathematics Education*, Mexico, Vol. 3, p. 129-136.

Pinxten, R.: 1988, *Towards a Navajo Indian Geometry* Ghent:Boeken Reeks.

Pirie, S. E. B. and Kieran, T. E.: 1991, *A Dynamic Theory of Mathematica Understanding - some features and implications*, paper presented at American Educational Research Association, Chicago.

Rotman, B.: 1988, Towards a Semiotics of Mathematics *Semiotica* Vol. 72, 1/2, p. 1-35.

Said, E.: 1974, An Ethics of Language, Review of M. Foucault 1972 "The Archeology of Knowledge" and "The Discourse on Language", Pantheon Books, New York, in *Diacritics*, Summer, p. 28-37.

Scott-Hodgetts, R. and Lerman, S.: 1990, Psychological/Philosophical aspects of Mathematical activity: Does theory influence practice?, In *Proceedings of Fourteenth Annual Conference of the International Group for the Psychology of Mathematics Education*, Mexico, Vol. 1, p. 199-206

Shankar, S. G.: 1987, *Wittgenstein and the Turning Point in the Philosophy of Mathematics* Croom Helm.

Steffe, L. A. and Cobb, P.: 1988, *Construction of Arithmetical Meanings and Strategies* New York:Springer-Verlag.

Sutherland, R.: 1992, Some Unanswered Research Questions on the Teaching and Learning of Algebra, *For The Learning of Mathematics*, Vol. 11, No. 3.

van Hiele, P. M.:.1986, *Structure and Insight* Orlando, Fla:Academic Press.

von Glasersfeld, E.: 1983, Learning as a Constructive Activity *Proceedings of the Fifth Annual Meeting of the North American Chapter of the International Group for the Psychology of Mathematics Education*, Montreal, Vol. 1, p. 41-69.

von Glasersfeld, E.: 1991, Knowing Without Metaphysics: Aspects of the Radical Constructivist Position, In F. Steier (Ed.) *Research and Reflexivity*, London:Sage, p. 12-29.

von Glasersfeld E.: 1992, *A Radical Constructivist View of Basic Mathematical Concpts*, Paper presented at Topic Group 16, Seventh International Congress on Mathematical Education, Quebec.

Vygotsky, L.: 1986, *Thought and Language* edited by A. Kozulin, MIT Press, Cambridge, Massachusetts.
Walkerdine, V. & Girls And Maths Unit: 1989, *Counting Girls Out*, London:Virago.
Walkerdine, V.: 1988, *The Mastery of Reason*, London:Routledge.
Weedon, C.: 1987, *Poststructuralist Theory and Feminist Practice*, Oxford:Blackwell.
Wittgenstein, L.: 1958, *Philosophical Investigations*, Oxford:Blackwell.
Wittgenstein, L.: 1974, *Philosophical Grammar*, Oxford:Blackwell.

Stephen Lerman,
Centre for Mathematics Education,
School of Computing,
South Bank University,
103 Borough Road,
London SE1 0AA U.K.

Mathematics Education Library

Managing Editor: A.J. Bishop, Melbourne, Australia

1. H. Freudenthal: *Didactical Phenomenology of Mathematical Structures.* 1983 ISBN 90-277-1535-1; Pb 90-277-2261-7
2. B. Christiansen, A. G. Howson and M. Otte (eds.): *Perspectives on Mathematics Education.* Papers submitted by Members of the Bacomet Group. 1986. ISBN 90-277-1929-2; Pb 90-277-2118-1
3. A. Treffers: *Three Dimensions.* A Model of Goal and Theory Description in Mathematics Instruction – The Wiskobas Project. 1987 ISBN 90-277-2165-3
4. S. Mellin-Olsen: *The Politics of Mathematics Education.* 1987 ISBN 90-277-2350-8
5. E. Fischbein: *Intuition in Science and Mathematics.* An Educational Approach. 1987 ISBN 90-277-2506-3
6. A.J. Bishop: *Mathematical Enculturation.* A Cultural Perspective on Mathematics Education. 1988 ISBN 90-277-2646-9; Pb (1991) 0-7923-1270-8
7. E. von Glasersfeld (ed.): *Radical Constructivism in Mathematics Education.* 1991 ISBN 0-7923-1257-0
8. L. Streefland: *Fractions in Realistic Mathematics Education.* A Paradigm of Developmental Research. 1991 ISBN 0-7923-1282-1
9. H. Freudenthal: *Revisiting Mathematics Education.* China Lectures. 1991 ISBN 0-7923-1299-6
10. A.J. Bishop, S. Mellin-Olsen and J. van Dormolen (eds.): *Mathematical Knowledge: Its Growth Through Teaching.* 1991 ISBN 0-7923-1344-5
11. D. Tall (ed.): *Advanced Mathematical Thinking.* 1991 ISBN 0-7923-1456-5
12. R. Kapadia and M. Borovcnik (eds.): *Chance Encounters: Probability in Education.* 1991 ISBN 0-7923-1474-3
13. R. Biehler, R.W. Scholz, R. Sträßer and B. Winkelmann (eds.): *Didactics of Mathematics as a Scientific Discipline.* 1994 ISBN 0-7923-2613-X
14. S. Lerman (ed.): *Cultural Perspectives on the Mathematics Classroom.* 1994 ISBN 0-7923-2931-7

KLUWER ACADEMIC PUBLISHERS – DORDRECHT / BOSTON / LONDON

MIX
Papier aus verantwortungsvollen Quellen
Paper from responsible sources
FSC® C105338

If you have any concerns about our products,
you can contact us on
ProductSafety@springernature.com

In case Publisher is established outside the EU,
the EU authorized representative is:
Springer Nature Customer Service Center GmbH
Europaplatz 3, 69115 Heidelberg, Germany

Printed by Libri Plureos GmbH
in Hamburg, Germany